DARWIN'S BLIND SPOT

Books by Frank Ryan

The Forgotten Plague

Virus X

Taking Care of Harry

Darwin's Blind Spot

DARWIN'S BLIND SPOT

Evolution Beyond Natural Selection

Frank Ryan

Houghton Mifflin Company
BOSTON • NEW YORK
2002

For my sister, Mary,

and my brother, Tony

Visit our Web site: www.houghtonmifflinbooks.com.

Library of Congress Cataloging-in-Publication Data
Ryan, Frank, date
Darwin's Blind Spot : evolution beyond natural
selection / Frank Ryan.
p. cm.
Includes bibliographical references (p.).
ISBN 0-618-11812-8
1. Evolution (Biology) 2. Natural selection.
I. Title.
QH366.2.R93 2002
576.8 — dc21 2002032284

Printed in the United States of America

Book design by Victoria Hartman

QUM 10 9 8 7 6 5 4 3 2 1

· *Acknowledgments* ·

I AM INDEBTED to Susan Canavan and David Wilson, who commissioned this book for the United States and the United Kingdom, respectively. Without their enthusiastic support it never would have seen the light of day. My special thanks to my editor, Laura van Dam; my manuscript editor, Peg Anderson; and my agent, Bill Hamilton, whose encouragement over the long years has been critical to my writing books of this nature.

The scientists were unbelievably courteous and generous with their time and assistance, not only sharing the joys and labors of lifetimes but also allowing me access to manuscripts of papers and books not yet published. I have listed the names of many new and distinguished friends and a few equally distinguished of long standing in alphabetical order, for their respective contributions will be abundantly obvious from the text and references: Redouan Bshary, Michael A. Crawford, Angela E. Douglas, Michael W. Gray, Daniel H. Janzen, Richard Law, Joshua Lederberg, James Lovelock, John McCoy, Lynn Margulis, Trevor A. Norton, Kris A. Pirozynski, David J. Read, Werner Reisser, Jan Sapp, Ruth Shady, F. Willem Vas Dias, Luis P. Villarreal, Donald I. Williamson, and Terry Yates.

Finally, a special thanks to Mike McEntegart and Chris Potter, who taught me microbiology and virology, respectively, a long time ago; and to Brian Jones, who oversaw my humble researches into the immune response of mammals, including myself, to the introduction of a virus known as ϕX174 into their bloodstream.

Contents

PART II
THE WEAVE OF LIFE

Why should all the parts and organs of many independent beings, each supposed to have been separately created for its proper place in nature, be so invariably linked together by graduated steps? Why should not Nature have taken a leap from structure to structure? On the theory of natural selection, we can clearly understand why she should not; for natural selection can act only by taking advantage of slight successive variations; she can never take a leap, but must advance by the shortest and slowest steps.

— CHARLES DARWIN,
The Origin of Species by Means of Natural Selection

· *Introduction* ·

A MYSTERY OF NATURE

> Our universe is a sorry little affair unless it has in it something for every age to investigate . . . Nature does not reveal her mysteries once and for all.
>
> — SENECA, *Naturales Questiones,* Book 7

THE FIRST-CENTURY ROMAN philosopher and statesman Lucius Annaeus Seneca had the misfortune of serving as the emperor Nero's adviser. At sixty-nine, he was returning to Rome from retirement in Campania when he learned of the death sentence that Nero had imposed on him. Seneca met his fate calmly, embracing his wife and friends and asking them to moderate their grief through reflection on the lessons of philosophy. His contribution to civilization remains relevant today, especially his insight into the mysteries of nature.

Two millennia later, Kwang Jeon was privileged to investigate such a mystery. His investigation came about indirectly, as so many important discoveries do. Though Jeon would be much too modest to claim so, the mystery he unraveled lies at the very heart of the evolution of life on Earth.

Now a distinguished research scientist in the department of biochemistry at the University of Tennessee at Knoxville, Jeon one day in 1966, while studying the humble amoeba at the State University of New York at Buffalo, experienced an unexpected calamity: his cultures of amoebae had been struck down by a plague. When he investigated, he found that they had been infected with an unknown strain of a bacterium later called the X-bacterium. He had no more idea where the infection had come from than the inhabitants of Europe had of the origins of the Black Death in the Middle Ages. "It just arrived, out of the blue. And suddenly, all of my amoebae began to die."

To find out if this bacterium was really causing the plague, Jeon tried infecting a few amoebae with the bacteria; the amoebae died. In the everyday world of clinical microbiology, that observation would have been enough: the X-bacterium would have been seen as a parasite to be eliminated. Jeon would have eradicated all the infected amoebae and sterilized his laboratory before starting the weary process of rebuilding his cultures.

The great Louis Pasteur once made a perceptive statement about the role of serendipity in scientific discovery: "Chance," he declared, "favors only the prepared mind." In Jeon, a combination of personality and circumstance ensured that chance had favored such a mind.

Korean-born Kwang Jeon never intended to become a biologist. At junior high school he had wanted to become a doctor of medicine: "In my young mind, I felt that I wanted to help others and do research on illnesses." The Korean War put an end to that ambition. But then he had a stroke of luck. In 1961, having just completed his master's degree in Seoul, he was adopted by the British Council and sent to London, where he studied zoology at King's College. After completing his doctorate, he became involved in research for the first time. Once again he ended up on a path he had not originally intended to follow. He was interested in embryology, but several people in the Department of Zoology happened to be studying the amoeba and he rather reluctantly joined their research program. His imagination was quickly captured. "When for the first time I actually saw amoebae moving under the microscope, I was fascinated." Just so does the thrill of discovery often begin for a scientist.

Most of us are familiar with the amoeba from high school biology, though we soon forget about it, assuming it has no relevance to our lives. This single-celled creature that lives in the mud of freshwater streams and ponds is about a fiftieth of an inch across and consists of a nucleus, which contains the DNA, surrounded by cytoplasm. We might even recall from biology class that the amoeba moves by pushing out blunt, fingerlike processes known as pseudopods. The young and curious Kwang Jeon asked himself how a cell with no limbs or obvious skeletal structures could engage in purposeful locomotion. "I watched, in a perfect stillness, how they put out their pseudopods to move."

He put amoebae into an environment with another creature, the green hydra. From the human perspective, hydras are minuscule, but they are predatory giants compared with amoebae. They also have a distinctive manner of feeding, pouncing on smaller life forms that come up close, stinging them with poisonous tentacles prior to devouring them. "I was expecting them to make a meal of my amoebae. But what actually happened was the hydra was gobbled up by the amoebae."

Jeon's love affair with the amoeba had begun.

One of the commonest amoebae in the world is *Amoeba proteus*. One strain of this species, known as the "D" strain, was discovered in Scotland in the early 1950s and subsequently found its way into research laboratories. Biologists were interested in the D strain because its tissues were known to contain some curious passengers: living particles of unknown origin with a striking resemblance to bacteria. While working in London, Jeon became very interested in the D strain of *Amoeba proteus*.

Like the hydra, the amoeba is a predator of even smaller creatures, such as *Colpidium*, enfolding them, together with a drop of water, within its pseudopodia. This process is known as phagocytosis. But the amoeba is prey to infection by certain bacteria, which it ingests in much the same way. The bacteria are resistant to the amoeba's digestive processes, and they infect it, causing serious ill health and sometimes death.

Some years later, Kwang Jeon moved to Buffalo, taking his beloved amoebae with him, to study under Jim Danielli, a world-renowned theoretical biologist who was interested in the phenomenon of cytoplasmic inheritance. The cytoplasm is the outer zone of the cell, separated from the nucleus by a double membrane. In the 1940s, a small number of scientists, including Danielli, began to doubt that all the hereditary programming was confined to the nucleus. To conventional biologists such doubt was outrageous. Since the close of the nineteenth century, biologists had been convinced that the nucleus was the sole repository of hereditary factors, so that any notion of cytoplasmic heredity seemed almost blasphemous. But those who took the idea seriously were very interested in finding out what actually happened when a microbe, containing its own genetic information, entered the cytoplasm of a cell whose hereditary information was supposedly confined within the nucleus.

Jeon's perspective on the plague that wiped out his amoebae was radically different from what might have been expected of a conventional microbiologist. Examining the lethal epidemic, he discovered that not all of the infected amoebae died. The precious few that survived did so even though their cytoplasm carried tens of thousands of living X-bacteria. It was clear that these few amoebae differed from all the others, possessing some inherited resistance to the plague bacillus. Intrigued, Jeon put aside his populations of uninfected amoebae, making sure they were protected from contamination, and began a new line of experiments, studying the interaction between the infected amoeba and the X-bacterium.

His experiments continued for many years, with some startling observations. After infecting an amoeba, the X-bacteria were resistant to the digestive enzymes that would normally devour them. Infection was followed by multiplication of the bacteria in such massive numbers that the host died, releasing large numbers of bacteria to infect others. This sequence explained both the lethality and means of spread of the plague.

But in the tiny minority of resistant amoebae, the process was very different. The bacteria took up permanent residence in the amoeba's cytoplasm, as if they had found a new home. And then the amoebae began to change. Newly infected ones — called xD amoebae — grew faster than those that had not been infected. They also seemed more fragile, being more vulnerable to starvation, overfeeding, and even minor temperature changes. They were so exquisitely sensitive to overcrowding that in normal colony densities the infected strain simply curled up and died. The hybrid life form might not have survived in nature because of this vulnerability.

Other changes in the xD amoebae were stranger still. The "genome" is the name scientists have given to the sum total of the genes that make up the heredity of any given species. For example, our human genome is made up of about 40,000 genes, parceled out in 46 chromosomes. At the time of Jeon's experiment, biologists thought that all of this genetic material was confined to the nucleus. Jeon wondered if the interaction between the amoeba and the bacteria was confined to the cytoplasm or whether there might be some nuclear component. Knowing that the amoeba's nucleus is remarkably tough, he placed two amoebae side by side and used a blunt probe to transplant the nucleus from one into the other. Normally the recipient amoeba would tolerate this transplantation very well. But when he transplanted a nucleus from an xD amoeba into a normal one, the grafted nucleus killed the recipient amoeba. This told Jeon that the infection was not affecting just the cytoplasm. It had changed the nucleus of the xD amoeba in some way that made it lethal to others.

Interaction with the amoeba also changed the bacterium. While initially up to 160,000 bacteria were found infecting a single amoeba, now, as some kind of equilibrium became established, the numbers of bacteria fell to about 45,000. Stranger still, if he removed the bacteria from the amoeba, they were no longer able to grow and reproduce in a laboratory culture. The bacteria could not survive outside the cytoplasm of their partner. At the same time, the host amoeba had become dependent on the bacterium for its survival. From an evolutionary perspective, something remarkable was taking place. Two utterly different species had melded into one, creating a new life form that was a hybrid of amoeba and bacterium. And the time frame was also interesting: the union was virtually instantaneous, although some further honing of the relationship continued afterward.

Some thirty-five years after it began, Dr. Jeon's experiment is still continuing, but already he has solved a little of the mystery. When I asked him if he had found any evidence for direct genome-to-genome interaction during the evolution of the hybrid, he replied: "We think, now, that we have a handle on some aspects of this question. For example, the bacteria somehow suppress a gene that would normally be essential for the amoeba."

Many genes work by coding for the manufacture of proteins that play an important role in the body's inner chemistry. In the hybrid amoebae an important enzyme was no longer being coded by nuclear genes, yet the enzyme was still in place and played a vital role in the hybrid's chemistry. In Jeon's words, "The enzyme must be coming from somewhere. Our feeling is that the bacterium is now supplying the gene for the enzyme."

In evolutionary terms the implications of this experiment are iconoclastic, differing radically from the theory proposed by Charles Darwin. Darwin believed that evolution proceeds by the gradual accumulation of small changes within individuals, governed by natural selection. Competition between individuals within a single species was the driving force. But what Jeon has observed is the union of two dissimilar species in a permanent living interaction. Their genomes, comprising thousands of genes that had evolved over a billennium of separate existence, have, in the evolutionary equivalent of the blink of an eye, melded into one. This is an example of an evolutionary mechanism known as "symbiosis," which is very different from Darwin's idea of natural selection. Today the overwhelming evidence suggests that this interactive pattern has played a formative, if largely unacknowledged, role in the origins and subsequent diversification of life on Earth.

Symbiosis complicates the unitary viewpoint taught in biology classes, but it brings a wonderful new perspective on life in general and on human society in particular. From the very beginning, evolutionary theory has been applied to many fields of human affairs, such as sociology, psychology and even politics. Such interpretations, viewed from a Darwinian perspective alone, lead to an excessive emphasis on competition and struggle. Most damaging of all, the social Darwinism of the first half of the twentieth century led directly to the horrors of eugenics. The rise, once more, of social Darwinism is therefore a source of worry to many scientists, philosophers, and sociologists. Recently, some evolutionary psychologists have gone so far as to suggest that rape may be a natural behavior. A broader understanding of evolution, taking into account not only interactions between species but also cooperation within our human species, would introduce some sense of balance into our understanding of these highly controversial aspects of human societal and psychosexual behavior.

PART I

Controversies

The Struggle for Recognition

It is the customary fate of new truths to begin
as heresies and to end as superstitions.

— THOMAS HENRY HUXLEY, *Science and Culture*

· 1 ·

THE ORIGINS OF LIFE

> We are so obsessed with finding other life forms, and with how life originated, it's as if life needs to seek itself out. That is manifest in our thinking.
>
> — PROFESSOR CAROLYN PORCO, the Lunar and Planetary Laboratory, Arizona

ALL AROUND US, in the most ordinary aspects of our existence, is the weave of life, so familiar we easily ignore its beauty.

Life is the ultimate mystery, from the diversity of species that inhabit the Earth to the labyrinthine complexity of the cells that make it possible. But how did such wonders come to be? This question, perhaps the most fundamental one that humanity has ever asked, is what evolution is all about. It is hardly surprising that many people still believe that life could have been created only by an omniscient and omnipotent God.

On March 31, 1998, I attended an auction at Sotheby's in London of the collected memorabilia of George Cosmatos. I was particularly interested in lot 318, consisting of a faded sheet of paper, once the dedication page of a book, which contained just twenty-two words handwritten in Latin. This piece of paper, the size of a page in a paperback novel, sold for no less than $42,000, which, if you added the buyer's premium, amounted to a staggering $46,000. These are the words written on the paper:

Numero pondere et mensura Deus omnia condidit
Hoc symbolum suum honoris et benevolentiae
Gratia Dignissimo Doctissimoque huus Albi
Possessori posuit

In English it reads, "Number, weight and measure, God created all these things. I have placed this, my motto, for the honor and best wishes of the most worthy and learned possessor of this book."

The clue to the high level of interest in lot 318 lay in the signature below the motto: that of none other than the great English scientist Sir Isaac Newton. To Newton, God had created all life, with humanity at its apogee: we were no less than the image made flesh of our divine maker. In Newton's day, most of his fellow countrymen, including scientists, believed that the Bible was the revealed word of an all-knowing God, who had created the Earth as described in the Book of Genesis. Even today some scientists still believe in the essence of this creationist theory of the origins of life.

The greatest upheaval in the history of biology began very modestly when, on July 1, 1858, a paper on a new theory of evolution was first read aloud to "thirty-odd nonplussed fellows" at a meeting of the Linnaean Society of London. But this no more than lit the fuse for the time bomb that duly exploded a year later when, in *The Origin of Species by Means of Natural Selection,* the English naturalist Charles Darwin expanded on the paper to propose a scientific basis for the origins of life. Darwin was not alone in proposing this theory; Alfred Russel Wallace shared its discovery with him. Subsequently Wallace came to disagree with Darwin in a number of ways, particularly the application of evolution to humanity. His mind wandered to teleological explanations and dalliances with spiritualism and even phrenology. Consequently, Darwin's interpretation assumed center stage, dominating every branch of evolutionary theory.

More than a century later, we have difficulty imagining the ripples his theory caused in the relatively tranquil pond of accepted belief at the time. Darwin insisted that life had not been created in six days, as stated in the Book of Genesis, but that it had come about through a process of gradual change under the influence of "selection" by nature. Such a revolutionary thought went far beyond the world of biology. Ernst Mayr, the Alexander Agassiz Professor of Zoology Emeritus at Harvard University, did not exaggerate when, in *One Long Argument,* he eulogized *The Origin* as "the book that shook the world." Out went determinism, based on creationist theology; in came concepts such as probability and chance. Almost at once our self-centered conception of our own existence was overthrown. Inevitably, from its inception, Darwin's theory of evolution encountered fierce resistance.

Darwin's ideas ran counter not only to religious faith but also to the prevailing scientific dogmas. His hypothesis was attacked and vilified by representatives of the established churches and, with equal force, by contemporary philosophers and many of his fellow scientists. The great English astronomer Sir John Herschel, regarded as the foremost physicist of his day, dismissed the probabilistic nature of natural selection as "higgledy-piggledy," while at Harvard the zoologist Louis Agassiz dismissed it as "a scientific mistake, untrue in its facts, unscientific in its methods and mischievous in its tendency."

The problem with much of this counterargument to Darwinism was a flaw that one might call the "Procrustean stance," after the Greek myth of Procrustes, who invited his victims to sleep on his bed. If they were too short, he stretched their bodies on his rack to make them fit, and if they were too tall, he chopped off their feet. In the 1650s, the archbishop of Ireland, James Ussher, had dated the act of creation to 4004 B.C., based on a meticulous reading of the Bible. For two hundred years this chronology was accepted, and the history of life was made to fit Ussher's extrapolation — an example of the Procrustean stance. In looking to faith rather than logic, Ussher accepted the literal truth of the Bible without questioning it, assuming his conclusion from the very beginning.

Science, because it is based on logic rather than faith, cannot take such a Procrustean stance. I believe in science because it provides us with a rational system of beliefs based on human logic, backed up by experimental evidence and proof. I am not so arrogant as to claim that science is always right or that it can answer all questions. A sense of humility is as necessary to the scientist as it is to the truly devout. Few scientists today would argue that life was created in the forms we see now; rather, those forms are the result of a long and complex procession of changes we call "evolution." Scientists and nonscientists alike think they know pretty well what is meant by this term. It is the process by which life first began on Earth and by which that fragile glimmer, over the billions of years that followed, changed and diversified until it gave rise to every form of life, from the simple bacterium to the most complex and colorful plants and animals that have ever lived, including humans.

In time Darwin's theory became massively influential. Not only did it offer, for the first time, a logical way in which species could diverge from related species, it gave rise to an understanding of "common descent." Today most educated people assume the truth of natural selection, which has been only a little modified by the century and a half since Darwin first described it. Perhaps understandably, the pendulum has shifted to the opposite extreme: today all too many scientists assume that natural selection is the *only* mechanism of evolution. But Darwin himself was more modest in his conclusion. As Mayr makes clear, for decades after publication of *The Origin,* Darwin kept changing his mind about how species change and diversify. He was aware of inconsistencies difficult to explain on the basis of his original thinking. Where did the vast panoply of variation come from? Was the minor variation that arose from sexual mixing enough to give rise to new species?

From the outset, well-informed people had doubts about this. One of the leading contemporary botanists in America, Asa Gray, supported Darwin, yet he could not accept that natural selection was sufficient in itself to explain the evolution of life. "Natural selection is not the wind which propels the vessel, but the rudder which . . . shapes the course." For Gray the only logical explanation was that a divine creator supplied the necessary variation, from which natural selection could choose the fittest individuals.

Some other biologists were beginning to think along different lines.

· 2 ·

THE OTHER FORCE OF EVOLUTION

> In view of this central position of the problem of species and speciation . . . one would expect to find in *The Origin* a satisfactory and indeed authoritative treatment of the subject. This, curiously, one does not find.
>
> — ERNST MAYR, *One Long Argument*

WHEN BIOLOGISTS look closely at nature they cannot help but notice cooperative partnerships that do not comfortably fit with the competitive struggle that is central to Darwinian evolution.

The hermit crab finds its home in the vacated shell of a whelk or a mollusk. To protect its vulnerable hind parts, it curls backward into the shell, securing the entrance with its armored claws. One species of hermit crab carries a large pink anemone on top of its shell. Fish and octopuses like to feed on hermit crabs, but when they approach this species, the anemone shoots out its brilliantly colored tentacles, with their microscopic batteries of poisoned darts, and sting the potential predator, encouraging it to look elsewhere for its meal. This is a perfect example of living cooperation, since the anemone in turn feeds on the droppings and leftovers from the crab's meals. The crab and the anemone appear to recognize each other as partners by tuning in to individual chemical signals — the equivalent of a bloodhound's fine-tuned sense of smell. The relationship is so firmly established that when the growing crab has to find a bigger shell, it delicately detaches the anemone from the old one and transports it to their new home.

Coral reefs are replete with such partnerships. Off the coast of the Indonesian island of Sulawesi lives a beautiful shrimp, the bright yellow of a buttercup, that lives in partnership with a cream-and-purple-banded fish known as a gobi. The shrimp works hard all day turning over reef debris for food to feed them both, while the gobi watches out for the predatorial approach of scorpion fish that might eat the work-distracted shrimp. Certain anemones play host to clown fish, whose silver-striped bronze heads can be seen peering out of the deadly tentacles. Another crab cleans the surface of a sea cucumber and is rewarded with a sanctuary in the sea cucumber's bottom. Still other crabs live in a mutually supportive relationship with sponges that dissuade predators because of their evil taste. And in the Indian Ocean a type of boxer crab, with a luridly checkered coat of red, white, and black, carries pastel blue anemones in both of its claws. The crab advances on its territorial rivals, holding both claws out and wielding its weapons like living knuckle-dusters.

Many of these relationships are not merely functional but also spectacularly beautiful. Some 319 species of hummingbirds are widely distributed throughout the warmer parts of the Americas, with a variety of dazzling iridescence unmatched by any other birds. They live almost entirely on nectar, a dependence on flowers that has led to an amazing diversity of form and color. Specialized joints in their wings enable them to beat so fast they are practically invisible, their whirring motion producing the characteristic hum; this adaptation enables them to hover with pinpoint accuracy and balance in front of the appropriate flower. Their beaks are exceptionally long and shaped to fit into the flower head, while their tongues, which are even longer, reach down into the well of nectar at the bottom. One of the most beautiful of all hummingbirds is the violet sabrewing, widely distributed from Mexico to Panama, which has a curved bill that fits the columnia flower as accurately as a scimitar fits its scabbard. Every time the sabrewing visits a flower to drink, the columnia's stamens are positioned to dab pollen on exactly the right point of the bird's forehead, so that it fertilizes the next flower it visits. In this way, flower and bird are partners in an exchange of food for assistance with fertilization.

Partnerships like these have been a source of wonder since ancient times. In his *History,* Herodotus described how the plover was known to take leeches out of the mouths of crocodiles, noting that "the crocodile enjoys this and, as a result, takes care not to hurt the bird." Aristotle observed a similar relationship between a bivalve and a crustacean, and Cicero was sufficiently impressed to draw the moral that humans should learn from such friendships in nature. The Roman statesman and scholar was remarkably prescient, for these associations are more than mere colorful scenes in nature. They have important implications for the nutrition, health, and long-term survival of the interacting partners.

These relationships pose an enigma: how can creatures not conventionally attributed with "intelligence" behave in such complex fashion? As far as we know, when the hermit crab puts the anemone onto its back to carry it to their new home, the crab does not "think" about its actions — any more than the anemone pauses to consider whether riding on the back of the crab is in its best interests. Other hermit crabs of that species do the same, as do others of that anemone species. The behavioral patterns of both crab and anemone have been hardwired into the genes of the partners by evolutionary forces over long periods of time.

In my book *Virus X,* I introduced the simple concept of genomic intelligence to help explain this. Genomic intelligence allows us to see that a seemingly rational behavior pattern of any life form, from a virus to a hippopotamus, is really an instinctive mechanism governed by the creature's genetic makeup. Given the anthropomorphic loading of the term "intelligence," it was predictable that this concept would be misunderstood, and it was. Although some might argue that genomic intelligence is no more than genetic programming, the concept is actually more subtle and complex. While an intelligent programmer creates a very specific computer program, genomic intelligence is not quite so fixed: on the contrary, evolution depends on its being able to change through various mechanisms. Genomic intelligence explains how behavior of a limited complexity is written into the genomes of crabs and anemones, enabling them to cooperate with each other in their daily lives and thus improve their chances of survival.

To biologists, such intimate relationships in nature pose an evolutionary dilemma. Darwinian selection is based exclusively on the individual's struggle for survival in competition with others of its own species. Here we find a pattern of evolution based on behavioral interactions between entirely different species.

An alternative approach to evolution began in the late nineteenth century, when the Swiss botanist Simon Schwendener became curious about the nature of lichens, those familiar growths that spread slowly over gravestones and rocks, somewhat floral in outline and often beautifully colored. Nineteenth-century biologists found lichens baffling. Initially they were thought to be primitive plants but when Schwendener examined them under the microscope he saw what appeared to be a combination of two life forms, a captive alga ensnared by fungal threads, or hyphae.

Schwendener could not imagine the two organisms, alga and fungus, living together in a mutually supportive relationship: instead he saw a master–slave relationship in which the fungus imprisoned the hapless alga and drove it mercilessly to work for its benefit. But when Schwendener first reported these findings to his colleagues, in 1868, he was derided and opposed. At this time biologists believed in a very rigid classification of life, based on the system put forward in the mid-eighteenth century by the Swedish botanist Carolus Linnaeus (Carl von Linné). Linnaeus invented the concepts of species and genus, founding the first system of logical classification, or taxonomy, of life. In the Linnaean system each life form was assigned to a single species, and the dual nature of lichens threatened this logic. It is not difficult to see why Schwendener's fellow lichenists were appalled by his claims of a bizarre union between a "captive algal damsel and a tyrannical fungal master."

In 1994 Jan Sapp, professor of the history of biology at York University, in Ontario, published the book *Evolution by Association,* which has become a landmark in the history of the study of symbiosis, which is known as symbiology. In analyzing the formative history of this discipline, Sapp describes how "a neutral term was required for symbiosis that did not prejudge such relationships as parasitic." In 1877 a German botanist, Albert Bernhard Frank, coined the term *Symbiotismus* to cover all relationships in which two different species lived on or in one another. This concept of a living association between species interested Anton de Bary, who had set up the first two institutes of botany in Germany and was the editor of the journal *Botanische Zeitung.* In 1878, just a year after Frank's pioneering definition, de Bary redefined symbiosis as "the living together of differently named organisms," leaving the precise nature of the association vague enough to include parasitism, commensalism, and mutualism. He thereby made the first detailed scientific case for symbiosis as both a biological phenomenon and, implicitly, a major force in evolution. Posterity has forgotten Frank's priority, and de Bary is now acknowledged as having first discovered and extensively investigated symbiosis.

De Bary, Frank, and Schwendener faced formidable opposition from their more conservative colleagues. Nevertheless, news about the Swiss and German discoveries began to spread, and biologists in many countries became interested. In 1873 the zoologist Pierre-Joseph van Beneden had introduced the term "mutualism" during a lecture to the Royal Academy of Belgium. Three years later Beneden brought these interdependent relationships to popular notice in *Animal Parasites and Messmates,* in which he wrote about the intriguing dependency between pilot fish and sharks. He also drew attention to other well-known associations, such as the Egyptian plover that cleans the teeth of crocodiles, the crabs that live inside the shells of mollusks, and the crustaceans, called cirripeds, that hitch a ride on the backs of sharks and whales. Only four years after de Bary defined symbiosis, a Scottish biologist named Patrick Geddes wrote a key article in the British journal *Nature* on the subject of "animals containing chlorophyll."

Forty years earlier the German zoologist Max Schultze had demonstrated that chlorophyll, the green pigment associated with plants, was present in certain species of planarian worms. Other scientists had subsequently found chlorophyll in a range of animals, including the freshwater hydra, freshwater sponges, the common green sea anemone, and even a crustacean. Geddes was particularly interested in a group of marine organisms called radiolarians, most of which contained strange yellow inclusions. Most of his colleagues dismissed the inclusions as glands, but in 1871 a biologist named Cienkowski made what Geddes termed "a very remarkable contribution to our knowledge" in demonstrating that they were independent parasitic algae, capable of living on in amoeboid form after the death of the radiolarian. Other scientists found similar "parasitic" algae in many species of sea anemones. Nevertheless, confusion and controversy continued as to the exact nature of these curious bodies, so Geddes decided he would travel to Naples and investigate the mystery for himself.

Geddes was particularly interested in the so-called yellow anemone, *Anthea cereus,* which in reality was not yellow but green. Indeed, he described it as "a far more beautiful green" than the species of anemone he had worked with before. Proving unequivocally that the green color was derived from algae, Geddes was able to demonstrate that the algae produced oxygen inside the living tissues of the anemone.

"What," he asked, "is the physiological relationship of the plants and animal thus so curiously and intimately associated?" He found it hard to believe that what he was observing was nothing more than parasitism. "Everyone knows that the colorless cells in plants share the starch formed by the green cells; and it seems impossible to doubt that the endodermal cell of the radiolarian, which actually encloses the vegetable cell, must similarly profit by its labors." Geddes observed that when an aquarium of anemones was exposed to sunlight, the hitherto motionless creatures suddenly began to wave their tentacles about, as if stimulated by the oxygen in their tissues. Moreover, he was well aware that just as the algae were producing starch that might be useful for the anemone, the waste products of the anemone were "the first necessities of life for our alga." The parasitic interpretation made no sense. If the alga was merely a parasitic invader, it should weaken its host, yet *Anthea cereus* was one of the most successful of all anemones.

When very different life forms evolve over millennia in close proximity to one another, some will come to influence one another. As Geddes went on to explain in his article and in two subsequent books, such cooperation was commonplace in nature.

Throughout the remaining years of the nineteenth century, this theme was taken up in a number of studies in which the most unlikely partners were found "living together" in mutually dependent relationships. It was the innovative German botanist Albert Bernhard Frank who made the most important of these discoveries, one that would radically alter our understanding of plant evolution.

In the 1880s the government of Prussia was interested in the cultivation of truffles, which were relished as a food delicacy throughout Europe. The Department of Agriculture and Forestry commissioned Frank to conduct scientific studies on the occurrence and development of these fungi. Little was known about truffles other than that they were usually found on or under the forest floor. The few facts that Frank could glean about them were very curious. Biologists tended to think of fungi as living on dead and decaying material, yet truffles seemed to occur only in and around the roots of living trees. Frank was also aware of an observation made by a botanist named Reess, who had noticed a bizarre union between the spreading threads, or mycelia, of a fungus and the roots of pine trees. In Frank's words, "From the outset these facts caused me to question whether true truffles also establish a mycelial connection with the living roots of trees."

At that time biologists regarded fungi in much the same way as they did bacteria: they were parasites. Botanists assumed that any involvement of a living plant with a fungus must therefore be an infection. Frank was far more open-minded than his colleagues in his research on truffles. That these fungi appeared to surround the roots of forest trees suggested an intriguing relationship. In a scientific paper published in 1885, he wrote: "Certain tree species . . . quite regularly do not nourish themselves independently in the soil but establish a symbiosis with a fungal mycelium over their entire root system. This mycelium performs a wet nurse function and takes over the entire nourishment of the tree from the soil."

Frank's astonishing revelations were derided in botanical circles; he had to fight to get his subsequent papers published. His findings, if true, would overturn the cherished beliefs of many senior botanists, notably the highly respected Theodore Hartig, who had, some forty years earlier, been the first to notice the root "mantle" of pine trees. He had assumed that this "Hartig net" was part of the normal root and failed to recognize its fungal nature. He went to his grave refusing to believe in the symbiotic union of trees and fungi.

Frank, meanwhile, conducted many further explorations into this curious, and wonderful, cooperation between fungi and trees. When he examined the roots of such familiar species as oak, beech, hornbeam, hazel, and chestnut, he found that the roots were actually composed of two different elements. The core of the larger roots was derived from the tree, but a mantle of fungal hyphae capped the stunted, club-shaped ends of these true roots. In many cases the mantle completely enclosed the root, forming such a closely woven covering around the growing tip that not even a root hair could escape. In looking at the tree roots, one saw only fungus. What appeared to be fine, filamentous root hairs radiating out into the soil were actually fungal threads growing out of the mantle. Frank did not believe that this was a parasitic infection. How could every tree in the forest be so grotesquely infected when the trees appeared to be perfectly healthy? He was convinced that the two very different elements formed some intimate cooperation, a union of two different beings into a single morphological organ. Frank called this organ a mycorrhiza, from the Greek for "fungus-root."

As realization dawned that he had made a discovery of monumental importance, the excited biologist searched harder, extending his inquiries to every type of tree he could find. The more he looked for this curious partnership, the more he found it. Although the patterns of mycorrhizae varied somewhat from one species to another, all the forest trees he studied had these curious mantles of fungi around their roots.

Frank was aware that swellings called tubercles had been found in the roots of legumes and that some biologists believed the tubercles contained masses of symbiotic soil bacteria. The bacteria were thought to have the ability to fix nitrogen from the atmosphere, which then helped to nourish the plant. Frank had no doubt that the fungi that formed mycorrhizae with the roots of trees were similarly beneficial. Far from being a parasite attacking the trees, the fungus, Frank surmised, was attracted to the roots by a chemical especially secreted for the purpose. Once established, the fungus enlarged the root area of the tree, increasing its absorptive capacity 10,000- or even 100,000-fold. In effect, the fungus was the mouth of the tree, imbibing salts, water, and organic nitrogenous food from the humus, while the tree, in return, supplied the fungus with the carbohydrate it needed for energy.

Controversy raged over Frank's theories. Older botanists opposed him tooth and nail, but more open-minded biologists extended his discoveries, finding mycorrhizae in association with many other plants, including orchids.

Orchids are the most diverse family of all the plants, with more than 17,000 species. In the late nineteenth century orchids were a fashionable topic in botany and subjected to intense investigation. The new thinking about symbiosis provoked a flurry of interest and excitement, and by the 1890s, fungi had been found in symbiotic relationships with no fewer than 500 species of orchids. In time the symbiotic connection between fungus and orchid was found to be even closer and more interdependent than that of trees. The fungal symbiont actually penetrates the orchid roots and enters the living tissues, supplying every nutrient the plant needs, even carbon, which in every other plant is fixed by the photosynthesizing leaves. With delightful aptness, the French botanist Noël Bernard even showed that penetration of the orchid by the fungus was necessary for the seeds to germinate. He compared the action of the fungi on the orchid to that of spermatozoa on eggs.

The intimate cooperation between wholly different life forms — plants and fungi — is not only an amazing biological phenomenon but also a vitally important factor in the diversity of plant life on Earth. It should have been of enormous interest to evolutionary theorists, but few scientists were paying attention. In those formative years at the end of the nineteenth century, as the fundamental principles of biology were being hammered into place in laboratories around the world, Darwinian evolution took center stage. And as Darwinism, with its emphasis on competitive struggle, thrived, symbiosis, its cooperative alter ego, languished in the shadows, derided or dismissed as a novelty.

At this timely moment a Russian anarchist decided to throw his hat into the ring.

· 3 ·

FROM ANARCHY TO COOPERATION

> Somehow or other, life appeared on Earth. I don't give a damn if somebody put it here, whether it was a waste bin cast aside by some visiting spacecraft or whether God fiddled with the chemicals and started it. Somehow it started here.
>
> —James Lovelock,
> BBC2 documentary "Life and Earth"

Peter Kropotkin was one of those rare individuals whose contributions to enlightenment are so original and inspiring that they blaze into a social prominence far beyond the confines of their scientific discipline. Though he was a member of the Russian nobility, Kropotkin's anarchist views repeatedly landed him in prison. Yet through every vicissitude he retained his delightful sense of humor. How could one dislike this prince-turned-anarchist who, still feeling himself constrained by courtesy, described how he took a stand against royal patronage: "A month ago I was invited to a banquet of the Royal Geographical Society in London. The chairman proposed, 'The King!' Everybody arose and I alone remained seated. It was a painful moment. And I was thunder-struck when immediately afterwards the same chairman cried, 'Long live Prince Kropotkin.' And everybody, without exception, rose."

It is a tribute to this interesting polymath that today he is quoted not only by the symbiotic school of thinking but equally by the Darwinians.

Kropotkin's interest in biology began when, at the age of eighteen, he first read Darwin's *Origin of Species.* Abandoning the long-term military career his father had in mind for him, he adopted the short-term strategy of taking a commission in the mounted Cossacks of the Amur. This permitted him to organize an expedition to the unexplored vastness of Siberia, where he could follow his natural interests to his heart's content. Meanwhile he was beginning to adopt a socialist attitude to political reforms. Although the czar had promised widespread improvements, Russia was turning away from social reform. Kropotkin took it upon himself to write reports on the conditions in the prisons of Siberia. The authorities responded by sending him on a series of exploratory journeys into the mountainous borderlands between Siberia and China, where he relished the opportunity of experience as both a field geographer and a naturalist observer. The young zoologist J. S. Poliakov, who had become his friend, accompanied him in exploring the uncharted territory between the Lena River and the upper reaches of the Amur. Inspired by Darwin's theory of evolution, the two men were anxious to find evidence of the competition between animals that would help confirm it.

To their surprise, they found little such evidence.

> We looked vainly for the keen competition between animals of the same species . . . We saw plenty of adaptations for struggling, very often in common, against the adverse circumstances of climate, or against various enemies, and Poliakov wrote many a good page on the mutual dependency of carnivores, ruminants and rodents . . . We witnessed numbers of facts of mutual support, especially during the migration of birds and ruminants; but even in the Amur and Usuri regions, where animal life swarms in abundance, facts of real competition and struggle between higher animals of the same species came very seldom to my notice, though I eagerly searched for them.

On his return home the czarist police threw him into prison. He escaped, making his way to Britain, where he kept himself alive by writing notes and columns for the science magazine *Nature,* and for the *Times.* Traveling to France and Switzerland, he carried on with political agitation until he was once more arrested in France in December 1882. Imprisoned in the ancient St. Bernard's monastery at Clairvaux, he received from a sympathetic fellow scientist a copy of a lecture, entitled "On the Law of Mutual Aid," that had been delivered in January 1880 to a congress of naturalists in Russia. The lecture, by Karl Fyodorovich Kessler, a zoologist and dean of the University of St. Petersburg, confirmed Kropotkin's own field observations, posing the theory that "besides the law of mutual struggle there is in Nature the law of mutual aid, which, for the progressive evolution of the species, is far more important than the law of mutual contest." Kropotkin was inspired by this. Keeping faith with Darwin, he perceived in "mutual aid" a development of ideas that Darwin himself had expressed in *The Descent of Man.* "This suggestion seemed to me so correct and of so great an importance, that . . . I began to collect materials for further developing the idea."

Thus, when Thomas Henry Huxley in 1888 published the essay "The Struggle for Existence" in the magazine *The Nineteenth Century,* Kropotkin had gathered enough evidence to refute it.

Huxley's essay, in retrospect, makes for uncomfortable reading. The tone is one of unrelenting bleakness, laced with his characteristic razor-sharp wit; the subject matter wanders from his views of animal and human evolution to the desperate situation of the poor, including the working classes toiling in the factories of the industrialized world. While statesmen and captains of industry declaimed against the curse of war and extolled the blessedness of peace and the innocent beneficence of industry, yet "so long as natural man increases and multiplies without restraint, so long will peace and industry not only permit, but they will necessitate a struggle for existence as sharp as any that ever went on under the regime of war."

Although the rule of logic is of central importance to science, scientists themselves are humanly prejudiced. Huxley was no exception, any more than Kropotkin was. No doubt the latter saw in evolution the rationale for his anarchist beliefs, in which a true society (each member giving what he or she could give, each receiving whatever he or she needed) lived in perfect cooperation without the need for central government. With this vision of utopia, he was offended by Huxley's bleak extrapolations of Darwinism to human evolution:

> For thousands and thousands of years, before the origin of the oldest known civilizations, men were savages of a very low sort . . . They preyed upon things weaker or less cunning than themselves . . . so, among primitive men, the weakest and stupidest went to the wall, while the toughest and shrewdest, those who were best fitted to cope with their circumstances, but not the best in any other sense, survived. Life was a continual free fight . . . The Hobbesian war of each against all was the normal state of existence.

At the time of Huxley's paper, there was very little fossil evidence of early humans. Paleontologists were still searching for the missing link. But Huxley certainly would have known of the fossils of one partial skeleton of an early human, which had caused a sensation when first found, just three years before the publication of *The Origin,* in a cave in the Neander Valley, near Düsseldorf, Germany. An Irish anatomist, William King, had first identified the find as evidence for a new species of humanity, *Homo neanderthalensis,* or Neandertal man. Those fossils were thought to show that early man was brutish in his proportions. The first real scientific appraisal of the fossils was made in 1911 by Marcellin Boule at the Museum of Natural History in Paris, who ignored the Neandertal's large brain and concluded that the creature had but rudimentary intelligence. He stated that the Neandertals were more apelike than human, with slouching posture and awkward gait, a back-sloping brow and prognathous jaw, exceptionally large teeth and no chin.

Even today the epithet "Neandertal" is synonymous with dimwitted brutishness. But subsequent excavations of Neandertal graves have shown plants, perhaps flowers or herbs, placed in a man's grave during burial, indicating that his family and group had almost certainly shown rituals of tenderness or religious belief when they buried him.

Kropotkin countered Huxley's view with a series of influential articles in the same magazine that had published Huxley's paper. These articles were later gathered together in the book *Mutual Aid: A Factor in Evolution,* in which Kropotkin emphasized the importance of cooperation as opposed to aggression as the formative principle of evolution.

This was not a battle between symbiology and Darwinism. Kropotkin espoused Darwin's theory but believed that others had horribly distorted it. Indeed, Kropotkin quoted the combative Huxley's words to show how Darwin's vision had been twisted into an extreme parody: "From the point of view of the moralist, the animal world is on about the same level as a gladiators' show. The creatures are fairly well treated, and set to fight; whereby the strongest, the swiftest and the cunningest live to fight another day. The spectator has no need to turn his thumb down, as no quarter is given." Dismissing such unrelenting pessimism, Kropotkin described how the "numberless followers of Darwin" had reduced the notion of struggle for existence to its narrowest limits. "They came to conceive the animal world as a world of perpetual struggle among half-starved individuals, thirsting for one another's blood."

Kropotkin argued his case articulately and with much personal observation from his own researches. Referring to the underpopulation of life forms that was the distinctive feature of the immense plains of northern Asia, he explained: "I conceived since then serious doubts — which subsequent study has only confirmed — as to the reality of that fearful competition for food and life within each species, which was an article of faith with most Darwinians, and, consequently, as to the dominant part which this sort of competition was supposed to play in the evolution of new species." Kropotkin amassed a determined case for cooperation at all levels of nature, whether between members of a single species or between very different species. In fact, however much he claimed to be a believer in Darwin's theory, his championing of cooperation ran contrary to the ethos of Darwinism at that time.

In 1798 the English cleric Thomas Robert Malthus had written his *Essay on the Principle of Population,* which argued that society should take a laissez-faire approach to human suffering. Malthus contended that populations had a natural tendency to increase beyond any potential supply of food. He claimed that food supplies could increase only arithmetically while populations burgeoned in a geometric fashion and that an uncontrolled population expansion must lead to disaster unless there were "positive checks" in the form of disease, famine, and war. He included in his "positive checks" the general struggle among individuals and classes by which the weakest go to the wall. If order and balance were to be maintained, he argued, these processes should not be tampered with. Malthus's views had greatly influenced Darwin's thinking, and even more so that of his more extreme followers. Kropotkin's belief in cooperation as a force of life offered a counterbalance to the Malthusian view of the struggle for existence.

In particular, Kropotkin opposed the extrapolation of that view to human affairs, or social Darwinism, which was promoted by such respected figures as Spencer and Huxley. The social Darwinists set out to prove that although humanity, through its higher intelligence and knowledge, might mitigate the harshness of the struggle for survival, nevertheless the pitting of human against human obeyed a relentless law of nature. Kropotkin disagreed. "This view, however, I could not accept, because I was persuaded that to admit a pitiless inner war for life within each species, and to see in that war a condition of progress, was to admit something which not only had not yet been proved, but also lacked confirmation from direct observation."

In this perception Kropotkin was not entirely alone. In 1844, some fifteen years before the publication of *The Origin*, the German socialist philosopher Friedrich Engels had published a treatise on economics in which he dismissed Malthus's theory on the grounds that it offered no convincing proof for the assumption of an imbalance between human population increase and food supply. Later on Alfred Russel Wallace, the codiscoverer of natural selection, also pointed out that Malthus had ignored the effects of political structure and social relationships. A member of the "established church," Malthus spoke from the same background of privilege as Darwin.

In *Mutual Aid*, Kropotkin's became the clearest voice raised in considered rebuttal of Malthus, Huxley, and Herbert Spencer. Beginning with evidence for cooperation among animals, he expanded this view to cooperation as an alternative strategy to competition in human society, declaring: "The mutual-aid tendency in man has so remote an origin, and is so deeply interwoven with all the past evolution of the human race, that it has been maintained by mankind up to the present time, notwithstanding all vicissitudes of history."

Tireless in his advocacy of such views, Kropotkin traveled widely, with two trips to North America in 1897 and 1901, lecturing on the importance of mutual aid in any interpretation of social biology. He believed that his was the true interpretation of Darwin's vision, for he accepted the importance of natural selection, and even the role of competition within and between species, questioning only its extreme Malthusian interpretation. But was Kropotkin's interpretation really closer to Darwin's than that of his bitter adversary Huxley?

Given his close friendship and long-standing relationship with Darwin, Huxley was far better placed to interpret Darwin than Kropotkin could ever be. Almost a century later, Lee Dugatkin, professor of biology at the University of Louisville, explained that Kropotkin was unable to see that Darwinians have difficulty in believing that individuals behave altruistically for the good of the larger group. Kropotkin believed that humans were naturally cooperative and that crime and violence were the results of governments that entrenched inequality; but as Dugatkin stated, "The literature on every sort of noncooperative act imaginable suggests that this view is naive — nice in principle, wrong in fact."

However naive he may have been in his political views, in his grasp of the universal principle of cooperation Kropotkin was not mistaken. The failure of his contemporaries to take into account the balancing ethos of cooperation in relation to competitive struggle was to have terrible human consequences.

· 4 ·

THE PRICE OF FAILURE

> But it is quite certain that no existing democratic government would go as far as we eugenicists think right in the direction of limiting the liberty of the subject for the sake of the racial qualities of future generations.
>
> — LEONARD DARWIN,
> Cambridge University Eugenics Society lecture, 1912

TO UNDERSTAND the reasons for the failure of Kropotkin's views to be accepted, one needs to look at the nature of society and the prevailing climate of belief in the nineteenth century and earlier years of the twentieth century.

Malthusian laissez-faire attitudes that would seem brutal today were accepted in nineteenth-century British society. Class divisions were much more clearcut than they are today, and at the bottom of the social structure was a teeming underclass, largely but not exclusively urban, uneducated, and teetering on destitution. For men, voting rights depended on a certain level of wealth and property; women, regardless of class or education, had no right to vote. The great centers of learning, Oxford and Cambridge, were the exclusive fiefdoms of the established church. A professor of geology had either to be a clergyman, or, as in the case of Charles Lyell, whose geological theories had influenced Darwin, had to be vetted for religious orthodoxy by the archbishop of Canterbury and the bishops of London and Llandaff.

In many ways Britain was probably the most enlightened of the "civilized" Western countries. When Darwin published *The Origin* in 1859, slavery was still legal in the United States and it would take a civil war to bring it to a close. In this zeitgeist the more sympathetic and cooperative ethos of symbiosis struggled for critical recognition against the competitive theory of Darwinism. Although Darwinism challenged the creationist viewpoint of Christianity, it posed no threat to societal mores. On the contrary, it espoused and validated the status quo. Class divisions, like wars of conquest between countries, were seen as natural consequences of the evolutionary struggle. To understand how evolutionary theory interacted with social belief, it is necessary to consider how ideas spread.

In his pioneering book *The Selfish Gene,* Richard Dawkins, the Charles Simonyi Professor of Public Understanding of Science at Oxford, invented the highly relevant concept of the meme, arguing that most of what is unusual about man can be summed up in one word: "culture." Dawkins saw cultural transmission as analogous to genetic transmission and thus a mechanism of evolution in its own right. Language in particular can evolve much faster than organisms do, and nowhere is cultural transmission more important than in the form of "communicable" ideas. To describe the phenomenon of contagious ideas, he took the Greek word *mimeme,* for imitation, and reduced it to "meme," simply because it sounded like "gene." Memes govern people's behavior and the evolution of social belief. Once "infected," people will incorporate the meme into their lives, as a central doctrine of truth.

Scientists are not immune to memes, which also play an enormous part in the communication of scientific messages and, ultimately, in their acceptance into popular consciousness. And for scientist and nonscientist alike, no meme is more easily accepted than one that confirms existing prejudices.

At the time *The Origin* was published, imperialism was the dominant political and social ethos in Europe. A nation was defined as "one that could defend its own boundaries," and the Britain in which Darwin lived as a comfortable country squire was the greatest imperial power since the Roman Empire. Contemporary Darwinism was in perfect harmony with imperialism, which was seen as the national expression of the evolutionary paradigm, the fittest nation dominating all others through the quality of its culture and the struggle of its armed forces. What was the empire but the just reward for quality and struggle! One of the most important arenas in which evolutionary memes gain power is, of course, political philosophy. And while Darwin eschewed any extension of his views to politics, his successors had no hesitation in carrying his ideas into more controversial arenas. By the late nineteenth and early twentieth century, Darwinian evolutionary theory was being applied to the fields of education, law, philosophy, behavioral psychology (including the sexual aspects), and politics. Translated into sociology, it became the social Darwinism that so offended Kropotkin.

As early as January 1860, Darwin's English contemporary Herbert Spencer published the essay "The Social Organism," in which he drew parallels between living organisms and societies at various levels of complexity. Spencer, who was privately educated rather than university trained, was more a philosopher than a biologist; and it was only after Darwin published his much more rigorously formulated hypothesis that Spencer was able to tie biological evolution to society, challenging Darwin's metaphor of "natural selection" with his harsher expression, "the survival of the fittest." That succinct phrase would change the history of the world.

For evolutionary biologists, terms such as "struggle," "competition," and "fitness" have precise biological meanings that differ from their everyday uses. The core Darwinian concept of fitness is not brute physical strength or stamina but reproductive success. Darwin made this abundantly clear in *The Origin:* "I use the term Struggle for Existence in a large and metaphorical sense, including dependence of one being on another, and including (which is more important) not only the life of the individual, but success in leaving progeny." This means that socially deprived parents who produce bigger surviving families are "fitter," in the evolutionary sense, than the most cultivated and brilliant members of society, if the latter produce few or no children. The eugenicists who turned this idea on its head, to favor the rich, cultivated, and brilliant, thus perverted biological theory for their own agenda.

Even Alfred Russel Wallace made the mistake of embracing Spencer's concept of "survival of the fittest" as a more objective conceptualization of natural selection, as indeed did the general public, for whom it seemed custom-made to fit all manner of prejudices and crackpot notions of ethnic and racial primacy.

Spencer himself stated explicitly that the rapid elimination of "unfit" individuals would benefit the human race, and thus the state should do nothing to relieve the conditions of the poor and needy. Quite apart from the cruelty implicit in this policy was the monumental arrogance of assuming that the only value of individual human beings was to further some culturally laden evolutionary purpose. For example, although his friendship with Mary Ann Evans, better known as the novelist George Eliot, might have persuaded Spencer that women had intelligence, he was convinced that their role was essentially reproductive; any attempt to educate them on a mass scale would overtax their brains and thus prove damaging to progress. A far more ruthless exponent of Malthusian thinking than Darwin himself, Spencer provided the defenders of laissez-faire capitalism with the intellectual justification to oppose state interference in market forces.

Spencer's world view was treated with suspicion by the more socially minded politicians in Britain, but it took a firm hold on the imagination of pioneering Americans. Peter Singer, DeCamp Professor of Bioethics at Princeton University, describes how the capitalist magnates used it to argue their case for the deregulation of industry in the Supreme Court. Andrew Carnegie used Spencerian thinking to justify the rat race of competition, "because it insures the survival of the fittest in every department." And John D. Rockefeller Jr. went so far as to use the analogy of the American Beauty rose, which could be brought to its full splendor and fragrance only by sacrificing the lesser buds. This wasn't an evil tendency in business but the working out of a law of nature and a law of God.

As Singer relates in his tiny jewel of a book, *A Darwinian Left,* "So often did the opponents of regulation appeal to Spencer, that Mr. Justice Holmes felt compelled . . . to point out that 'the Fourteenth Amendment does not enact Mr. Herbert Spencer's *Social Statics.*'"

Of course competition and struggle are normal elements of business, and Carnegie and Rockefeller were more than merely paradigms of capitalist selfish exploitation: both were liberal in temperament and generous philanthropists in practice. What their example reveals is the power of the Spencerian meme and the ease with which it so captured the public imagination. Spencer's editor Robert L. Carneiro openly admits that such Spencerian concepts have no basis in science. "Since they proclaim what ought (or ought not) to be done, they are tenets of political philosophy rather than scientific statements. As such one may freely reject them." But that is not how memes work. And while scientists today might object to this abuse of evolutionary theory, it was in fact scientists themselves, including some misguided representatives of the medical profession, who paved the way.

The notion that heredity can be controlled through breeding is ancient. Farmers and breeders have used breeding techniques for thousands of years to improve the yield of domesticated animals and plants for human use and consumption. The question inevitably arose whether humans could direct the evolution of their own species toward goals that might be defined as "good" or "desirable." References to such improvement of the human stock by deliberate selection appear in the Old Testament and Plato's *Republic.* For many members of the British middle class, the question acquired new urgency as well as a respectable scientific validity when, in *The Descent of Man,* Darwin explained how his evolutionary theories applied to humanity. As in all of Darwin's writings, animals were described with perception and sympathy, but he was somewhat less discerning about people. It is difficult for a twenty-first-century reader not to take offense at many of Darwin's assumptions about the superiority of Western, especially Anglo-Saxon, culture, and his patronizing references to so-called "inferior" peoples, which riddle the text. Equally objectionable are his views that medical and humanitarian programs were preserving the "unfit":

> We civilized men, on the other hand, do our utmost to check the process of elimination; we build asylums for the imbecile, the maimed, and the sick; we institute poor-laws; and our medical men exert their utmost skill to save the life of every one to the last moment. There is reason to believe that vaccination [of children] has preserved thousands, who from a weak constitution would formerly have succumbed to smallpox. Thus the weak members of civilized societies propagate their kind. No one who has attended the breeding of domestic animals will doubt that this must be highly injurious to the race of man. It is surprising how soon a want of care, or care wrongly directed, leads to the degeneration of a domestic race; but excepting in the case of man himself, hardly any one is so ignorant as to allow his worst animals to breed.

It seems incongruous to find such views being championed by a man who was described by his contemporaries as extremely modest and gentle, somebody who would go out of his way to avoid hurting the feelings of others. In mitigation, he added a paragraph explaining why we nevertheless feel we must give aid to the helpless. "Nor could we check our sympathy, even at the urging of hard reason, without deterioration in the noblest part of our nature." Unfortunately, urgings of "hard reason" would prove altogether more compelling to the eugenicists who followed Darwin than any appeal to the nobility in their nature.

The Descent of Man portrayed men as physically stronger and more intelligent than women, who are described as kinder and more sensitive. Thus, "Man is more courageous, pugnacious and energetic than woman, and has a more inventive genius." Such prejudicial views of women were far from abstract considerations. They were employed, ruthlessly and systematically, to keep women in their place. Darwin himself supported the view that women should be excluded from meetings of the Geological Society.

The effects of such bigoted thinking are illustrated by the example of Beatrix Potter, who had devoted years of her life to the classification of fungi, in particular to the dual nature of lichens. When, in 1890, she attempted to speak about her observations at the Linnaean

Society of London, she was barred from doing so. She suffered the further humiliation of having her uncle, the distinguished chemist Sir Henry Roscoe, present her findings at a meeting from which she was excluded. When similar obstacles were placed before her research in the British Museum and the Royal Botanical Gardens, she abandoned any hopes of a career in science, taking up instead the writing of her famous series of children's books.

Indeed, it seems at times that every prejudice of Victorian Britain finds expression in Darwin's book. Under the heading "Civilized Nations," Darwin included the opinions of his friend William R. Greg, a fervent phrenologist and brash apologist for the labor practices of the Lancashire mill owners. Greg was thus afforded a respectable platform for his prejudiced opinions of the entire Irish nation. Thus Darwin quoted Greg, word for word:

> The careless, squalid, unaspiring Irishman multiplies like rabbits: the frugal, foreseeing, self-respecting, ambitious Scot, stern in his morality, spiritual in his faith, sagacious and disciplined in his intelligence, passes his best years in struggle and celibacy, marries late, and leaves few behind him. Given a land originally peopled by a thousand Saxons and a thousand Celts — and in a dozen generations five-sixths of the population would be Celts, but five-sixths of the property, of the power, of the intellect, would belong to the one-sixth Saxons that remained.

For Greg, who understood the implications of fitness, the evolutionary outcome made no sense. "In the eternal 'struggle for existence,' it would be the inferior and *less* favored race that had prevailed — and prevailed by virtue not of its good qualities but of its faults."

But Darwin's readers could console themselves with the Malthusian argument that a provident nature had evolved checks to "this downward tendency." Darwin duly listed the checks that nature had placed in the path of too high a birthrate among the poor, in particular the high death rate of young mothers and of their children during the first five years of life in their crowded urban hovels. And so, in a painful progression throughout *The Descent,* the prejudices of upper-middle-class English pomposity, in the guise of established knowledge, condemn to inferiority an extensive list of groups, including all women, the Irish, the physically and mentally afflicted, the poor, and indigenous people in all corners of the globe. Blacks were stupid, inferior, and lazy, as evidenced by the numerous references to "savages" and discussions as to whether or not the races of man constituted distinct and separate species, in which the "inferior vitality of mulattoes" might be seen as possible evidence of the "specific distinctness of the parent races." Summarizing the "various checks" on undesirable breeding, Darwin went on to state that if these "do not prevent the reckless, the vicious and otherwise inferior members of society from increasing at a quicker rate than the better class of men, the nation will retrograde, as has too often occurred in the history of the world." "Progress," he cautioned, "is no invariable rule."

Huxley, though so often portrayed as Darwin's bulldog, expressed it differently:

> It needs no argument to prove that when the price of labor sinks below a certain point, the worker infallibly falls into that condition which the French emphatically call *la misère* — a word for which I do not think there is any exact English equivalent. It is a condition in which the food, warmth, and clothing which are necessary for the mere maintenance of the functions of the body in their normal state cannot be obtained; in which men, women, and children are forced to crowd into dens wherein decency is abolished and the most ordinary conditions of healthful existence are impossible of attainment; in which the pleasures within reach are reduced to bestiality and drunkenness; in which the pains accumulate at compound interest, in the shape of starvation, disease, stunted development, and moral degradation; in which the prospect of steady and honest industry is a life of unsuccessful battling with hunger, rounded by a pauper's grave.

It is curious to reflect that the passage above, so radically different in tone from *The Descent,* was part of the essay that Kropotkin, the anarchist, so strongly condemned. Concealed under the gruff affectations of misanthropy and antisentimentalism, Huxley is making a case for working-class improvement through technical education. "There is, perhaps, no more hopeful sign of progress amongst us, in the last half-century, than the steadily increasing devotion which has been and is directed to measures for promoting physical and moral welfare amongst the poorer classes." Although he may have overstated the bestial nature of the struggle for existence, he did not present such behavior as a desirable norm in human society. On the contrary, he stood, in his own words, "like a man" and implored enlightened humanity to do everything in its power to oppose such natural tendencies. "The ethical progress of society depends, not on imitating the cosmic process, still less in running away from it, but in combating it."

It is not altogether surprising that when Darwin's book on human evolution was first published, it elicited little of the fierce opposition he had encountered twelve years earlier with *The Origin*. Indeed, it was readily accepted by a society in perfect tune with its message, quickly selling a second edition that netted its author £1,500 — "a fine big sum," he boasted to Henrietta, his daughter, offering her £30 for her labors in helping him.

Sadly, though acknowledged both a genius and a genial man, Charles Darwin did not altogether rise above the prejudices of an English gentleman of his day.

Some years prior to publication of *The Descent,* Darwin's cousin Francis Galton became so convinced by the type of argument put forward by Greg that he decided humans must control their own evolutionary future. Galton pioneered the application of statistical methods to the study of heredity and founded the scientific discipline now known as genetics, a laudable achievement. Unfortunately, he is better remembered for having extrapolated Darwinism into a much more controversial arena.

In 1869 Galton published *Hereditary Genius,* in which he purported to show that "it would be quite practical to produce a highly gifted race by judicious marriages during several consecutive generations." In 1883 he took these theories a major step further when he published *Inquiries into Human Faculty*, coining there the term "eugenics," which he defined as the theory and practice of improving the human condition from a genetic point of view. There are some beneficial applications of eugenic thinking, for example our modern genetic counseling services, which allow informed decision making about various inherited risks without societal or professional pressure. However, the prevailing class misconceptions and racial and ethnic prejudices were extremely influential to eugenicist thinking. Poverty and criminality, for example, were linked, and the essential underlying cause was considered to be bad heredity.

In 1907 a twenty-one-year-old widowed Englishwoman, Sybil Gotto, with Galton's help, founded the Eugenics Education Society (later the Eugenics Society) in Britain, with the purpose of "promoting those agencies under social control which might improve the human race." As the historian Pauline M. H. Mazumdar makes clear, its interests were more social than biological. Galton defined the practical applications of eugenics as "the science of improving stock, which is by no means confined to questions of judicious mating, but which, especially in the case of man, takes cognizance of all influences that tend in however remote a degree to give the more suitable races or strains of blood a better chance of prevailing speedily over the less suitable than they otherwise would have." To promote these aims, he endowed a research fellowship and chair in eugenics at University College, London, both of which were, in the early years of the twentieth century, held by Karl Pearson, a brilliant mathematician who helped pioneer statistical methods in biology. Pearson never actually joined the Eugenics Education Society, preferring to keep his academic distance. Nevertheless, he took Galton's ideas to new extremes, believing that environment played little or no part in human mental or emotional qualities. Negative eugenics now had an evangelist.

Pearson accepted the earlier prejudices toward the poor as constitutionally inferior — they were "a breeding isolate on the margins of the human race" — and their negative morals and degenerate physical qualities made their high birthrate a threat to future civilization. Assuming that such moral and physical qualities were hereditary, the members of the Eugenics Education Society concluded "that if the prolific breeding of this class were not controlled, pauperism and its associated undesirable qualities must necessarily keep on increasing until the direction of evolution of the human race was reversed."

Although the Eugenics Society membership was not large, it included some of the most influential people in society. The Australian historian Lyndsay Farrall showed that 80 percent of the early members were listed in the *Dictionary of National Biography*, including the Reverend William R. Inge, future dean of St. Paul's Cathedral, many prominent biological and social scientists, and a small but significant number of doctors. Even the contribution of the mathematician Ronald A. Fisher, who would play a leading role in the "synthesis" revival of Darwinism in the 1930s, was tainted by his ardent commitment to wholesale eugenic practices, which he somehow reconciled with his equally ardent Christianity. Eugenics was not, however, supported by the British medical establishment as a whole and was never accepted into any British university medical curriculum.

Nonetheless, the committed members of the Eugenics Society found it possible to demonstrate, "in chilling manner," how pauperism was passed on from generation to generation. In lectures this process was illustrated with lengthy family pedigrees. Poverty was hereditary because the sons and daughters of the poor were themselves more likely to be poor. Included as evidence of degeneracy were medical conditions such as tuberculosis and epilepsy, which could be shown to run in families. Today we know that tuberculosis is not a hereditary disease but an infectious one, caused by the most lethal germ in history, *Mycobacterium tuberculosis.* Any child growing up in overcrowded, impoverished conditions in which a close relative was coughing up sputum containing the germ from morning to night was more likely than the rest of the population to contract the disease. Epilepsy, or a tendency to fits, affects 1 percent of the population, but its causative influences, often an otherwise minuscule scar in the brain, are hereditary only in a minority of sufferers. No serious person would consider Julius Caesar, Vincent van Gogh, Napoleon Bonaparte, Ludwig van Beethoven, or Charles Dickens as degenerates, although they are all believed to have suffered from epilepsy. Of course, for the eugenicists, there was a greater purpose in linking poverty, disease, and criminality to degenerative genetic causes: it meant that the "higher races" had a duty to make sure that future generations were not ruined by a progressive dilution with the more fecund "lower races."

In 1909, when Ronald Fisher went up to Cambridge to read mathematics, he came across Pearson's line of thinking. Two years later, Fisher helped found the Cambridge University branch of the Eugenics Society, one of many based in the British universities, including Oxford, Liverpool, Manchester, Belfast, and Glasgow, and extending to Australia and New Zealand. The meetings of the Cambridge branch, which were reported in detail in the *Cambridge Daily News,* capture the prevailing paranoia. At the inaugural launch, the Reverend Professor Inge expressed his fears that the "degenerates" of the urban proletariat "may cripple our civilization," as they had that of ancient Rome. The second public meeting was a "lantern" lecture by R. C. Punnett, professor of genetics at the university, who issued a call for action in words that are hard to reconcile with this otherwise avuncular historian of the early days of genetics: "We may object to the way in which God made some people; we may decide the world would be better without them. But it must be done calmly and without prejudice, in the clear light of reason, and not under the cloak of righteousness or of doing a thing that is pleasing to any but ourselves." Punnett was not calling for these degenerates to be put down like unwanted dogs but for the elimination of their kind by the enforced control of their future breeding.

The third public lecture to the Cambridge branch was delivered by none other than Darwin's son, Major Leonard Darwin, who later, as president of the Eugenics Education Society, strongly supported Fisher in his eugenics aims and his mathematical career. (So central was Fisher to the Cambridge branch that the eugenics meetings came to an end soon after he left the university.)

While many members of the British Eugenics Society supported charitable causes, promoting the education of the poor and fighting the scourge of alcoholism, it was inevitable that Pearson, as a respectable academic and a man of high intellectual ability, would give credibility to the many others who were committed to half-baked notions of the "hereditary taint" of inferior classes and races. It is disturbing to consider the impact of memes based on extremist application of Darwinian principles on the minds of doctors, teachers, employers, judges, and those on the control boards of the workhouses.

In 1912 the British Eugenics Society hosted the first International Congress of Eugenics, open to the public. A series of charts explained genetic inheritance alongside Galton's "Standard Scheme of Descent," and as a paradigm of the "inheritance of ability," one chart showed the family trees of the interrelated Darwins, Galtons, and Wedgwoods. The congress was attended by the faithful from many nations, including the American Breeder's Association, which put forward evidence that mental deficiency was hereditary. The largest section of the exhibition was taken up by the German contingent, which included such notables as Eugen Fisher, Max von Gruber, Ludwig Plate, and General von Bardeleben, the official genealogist to the German nobility.

By now the eugenicists in each nation had evolved their own notions of "ideal type." In Britain, for example, the pauper class was perceived as the greatest threat to civilization. In the United States, where social Darwinism became even more firmly entrenched than in Britain, wealth was equated with fitness and the poor were not to be assisted. Any attempt to change the domination of society by what came to be known much later as the WASP — or white Anglo-Saxon Protestant — would undermine the natural evolution of the human species. Theodore Roosevelt and Stephen B. Luce are judged to have supported a militaristic foreign policy on social Darwinian grounds, and the eugenicist Charles B. Davenport urged selective breeding to eliminate the physically and mentally infirm. When the American Eugenics Society was founded in 1926, its members claimed scientific corroboration for the idea that the white race was superior to all other races. Races, in this sense, were assumed to be "pure" lineages that had evolved in isolation from one another. Even within the white category, American eugenicists considered the Nordic white superior, naturally including themselves. And the undesirables with high birthrates were the feeble-minded and criminal classes, who were filling prisons and long-stay hospitals at such cost to the nation. In Germany the "undesirables" were identified as the psychotics and psychopaths filling up the mental asylums. In all such prejudicial assumptions, the one commonality was that the existing ruling classes had already proved their superiority.

Darwin would have been horrified to observe how his biological theory had been misused to support negative eugenic policies being put into place in many countries, with the aim of improving the species by identifying individuals and couples carrying "inferior" genes.

Those judged undesirable were to be prevented from reproducing, while people carrying "good" or "beneficial" genes were encouraged to have children. Inevitably, such decisions, put forward by some self-serving and socially powerful group determined to impose its opinion on more vulnerable groups, were subjective and controversial. In the United States, as in Britain and elsewhere, studies purported to show that entire families were degenerate, having inherited "bad genes." It was claimed that immigrants to the United States from southern and eastern Europe were innately inferior and given to criminality. Laws, such as the Immigration Act of 1924, sought to restrict immigrants from areas with suspect gene pools.

By 1931 enforced sterilization of undesirables had been introduced by law into twenty-seven states in America, and by 1935 similar laws had been passed in Denmark, Switzerland, Germany, Norway, and Sweden — all countries with substantial "Nordic" populations. Enforcement of these laws led to the involuntary sterilization of people judged to be "insane, mentally retarded or epileptic," as exemplified by William Faulkner's great novel *The Sound and the Fury.* Other groups that were similarly violated included "habitual criminals" and "sexual deviants." As a medical student in Britain in the late 1960s, I came across a poignant example of such eugenic madness: a woman who was condemned to roam the grounds of a large mental hospital after her frontal lobes had been severed from the rest of her brain as a treatment for "moral insanity." Her moral insanity consisted of having had two illegitimate children.

Social Darwinism took on a far greater potential for evil when it was adopted as policy by a ruthless dictatorship. In Germany, which was impoverished and demoralized following World War I, the eugenics meme found its most receptive audience. Adopted as gospel by the National Socialist party, the idea was seeded into the population through a combination of unrelenting propaganda and utter ruthlessness, culminating in the barbarities of programmed genocide. Again and again, in the writings of those advocating such inhumane policies, the reader is left in no doubt where the moral authority for such actions originated. Thus, citing Adolf Hitler, "It is the struggle for existence that produces the selection of the fittest." Regimes elsewhere were every bit as ruthless as Hitler's National Socialist Party, including the Communist party in the Soviet Union, Pol Pot's murderous policies in Cambodia, and the more recent genocidal tragedy in Rwanda. But these brutal killings were motivated more by political expediency and ethnic hatred than by eugenics. Even William H. Schneider, whose authoritative book on the French eugenicist movement tends to downplay the horrific implications, draws attention to the French eugenicist Georges Vacher de Lapouge, whose proposals "beginning in the 1880s, also contained an appreciation of the *possible far-reaching consequences* of Darwin's and Galton's ideas on heredity and human selection." Altogether revealing is the title of the book by the French science historian André Pichot: *La Société Pure: De Darwin à Hitler.*

· 5 ·

A CONFUSION OF TERMS

> Biologists ridiculed the idea that groups of organisms might gain a survival advantage over other groups because they shared some beneficial trait. Now that is changing. We are starting to understand that evolution happens on a variety of levels.
>
> — ecologist LYNN DICKS

EVEN AS THE EUGENICS SOCIETY was meeting in London, summer visitors to the coastal resort of Roscoff, in Brittany, might have observed a certain English gentleman behaving rather oddly. Walking at low tide across the golden beach, he passed scattered granite rocks before approaching the main brown belt of seaweed. Here he trod softly as he approached some dark green patches that glistened on the sand. Kneeling down carefully, he scooped up some of the green just before, astonishingly, the rest of it appeared to evaporate like magic from the warm surface of the sand. When, in response to the visitors' curiosity, he showed them his catch, they must have been exceedingly surprised.

The gentleman, Frederick Keeble, was a professor of botany at University College, Reading, and he was actually collecting not seaweed but minuscule worms of two species of the genus *Convoluta*. His listeners' dismay surely deepened as he explained that these two very lowly worms, no more than an eighth of an inch long, were not even segmented, as any decent high-class worm should be. But dismay invariably turned to interest as he explained why he was studying these creatures. The clue lay in their color, which he invited his listeners to examine with the aid of a magnifying glass. *Convoluta roscoffensis* was dark spinach green; *C. paradoxa* a brownish yellow. Indeed, these little "plant-animals," as Professor Keeble called them, were an odd composite of worm and plant. The otherwise transparent worms were packed with a brilliantly colored alga called *Platymonas,* which lived symbiotically within the tissues of the worms. The algae passed on the products of photosynthesis to the worms, which in turn donated their waste products to the algae.

Keeble might also have explained that nature abounded in such examples of symbiosis. Life was not merely an individual struggle for existence. The evidence for living interactions could be found everywhere if people cared to look.

Some ten years before Keeble began his studies at Roscoff, Patrick Geddes and J. Arthur Thompson had urged the world to take a more balanced view of biological reality. In the case of animals, hunger was only one of the driving forces of life. For all of life, procreation was of paramount importance, and it often involved mutual support between members of a species, with complex interactions between parents and offspring. Mutual support, they argued, was an important driving force of life. Recognition of symbiosis, the interspecies equivalent of mutual support, had been quietly growing. But obdurate Darwinians, if forced to acknowledge the existence of symbiosis, were determined to limit its importance.

In 1892 an American botanist, Roscoe Pound, speaking at a botanical seminar at the University of Nebraska, summarized the current status of symbiosis in the botanical world. His presentation so impressed his peers that it was subsequently published as a paper. After mentioning examples such as the relationship between humanity and wheat, the yucca moth and *Yucca,* and the familiar relationship in lichens between fungi and their algal hosts, he systematically downgraded the importance of symbiosis while advocating a Darwinian interpretation of all such phenomena. Pound, only twenty-three years old, no doubt represented the views of those who had taught him botany. But he was already director of the Nebraska state botanical survey and had recently discovered a rare fungus that was later named after him, *Roscopoundia.* Indeed, as Jan Sapp makes clear, Pound's opinions were important in a manner that went far beyond his youth.

"Who," he asked his seminar audience, "would not sympathize with those who derided de Bary's symbiotic descriptions of lichens!" Pound grudgingly accepted the reality of symbiosis in the plant kingdom, but he dismissed many purported examples of cooperation as "sheer exaggerations." Even if mutualism, for all "its seeming unreasonableness," could be shown to exist, it "does not exist in all lichens."

He reserved his utmost scorn for Albert Frank, the German botanist who had discovered mycorrhiza. "Frank asserts that certain species of algae have become so adapted to life in the lichens and so accustomed to it, that they have partially or wholly lost the power of independent growth. No examples of this, however, are certainly known." Again, referring to another of Frank's claims about lichens, Pound added, "It seems, like some other theories of Frank, which I shall have occasion to mention presently, if I may say so, decidedly 'fishy.'"

Microbes, Pound declared, were nothing more than parasites. Any perceived cooperation was a delusion resulting from superficial examination. The "*Rhizobia,* as Frank calls them," were another example of parasitic bacteria. As for Frank's mycorrhizal fungi, the notion that they cooperated with the trees or the orchids they infected was nothing more than "Frank's statements calculated to try our patience and credulity." Step by step, Pound undermined the extent and importance of Frank's discoveries, going so far as to side with Frank's colleague and most vehement detractor, Hartig — "a more sober and trustworthy writer than Frank" who "said the last word so far on Mycorrhiza in 1891." Although Hartig had admitted that some fungi were found around the roots of certain trees, "his conclusion is that they are of no use to the tree, and are probably injurious by taking nourishment properly belonging to the tree." Pound then gave a clue to his underlying thinking: "It would seem that they must do this, even were there mutualism between them and the roots — else why are they there?"

As we now know, Pound, like Hartig, was — one is tempted to say "Frankly" — mistaken.

The downgrading of symbiosis became a recurring pattern in the decades ahead. But in a single respect, Pound proved astute. As the aggressive-competitive ethos of Darwinism came to dominate evolutionary thinking, the semantic difference between symbiosis and mutualism became increasingly confused. Mutualism is a form of symbiosis in which benefit is conferred on all of the contributing partners. But symbiosis, in de Bary's original definition, is much wider in scope, including parasitic, commensal (that is, doing neither good nor harm to each other), and mutualistic relationships.

In the highly volatile politics of the second half of the nineteenth century, politicians and social reformers opposed to the aggressive-competitive ethos of Darwinism looked to mutualism in nature for an alternative philosophy. Inevitably, the concept became even more confused, as biologists and nonbiologists alike used the term to denote any loose kind of cooperation, not only between dissimilar species but also between members of the same species. Mutualistic ideas led to the growth of working-class organizations, such as trade unions and "friendly societies," which pooled resources to help one another. Such socialistic developments, in particular their links to trade unions, provoked increasing anxiety in the ruling classes. In time this trend would change the political spectrum in Britain, leading to the formation of the Labour Party. It will come as no surprise that it was also the left, and Marxist groups in particular, that put up the most effective challenge to the dominance of eugenics in the study of human genetics.

Mutualism, in the eyes of some people, became identified with notions of anticapitalism. It is little wonder that a biological phenomenon tinged with such social implications was no longer discussed in polite capitalistic society. Pound, perhaps, was a creature of his place and time.

Fortunately, not all Americans adopted a negative view of symbiosis. Albert Schneider, a lichenologist in Illinois, had a vision of symbiosis that went far beyond its shackling by Pound. In a seminal paper published in the first edition of *Minnesota Botanical Studies* in 1897, he opened with a challenging declaration: "All living organisms manifest a more or less intimate biological interdependence and relationship." Schneider saw symbiosis not as a rare event in an otherwise Darwinian world but as a commonplace biological situation; it was the very abundance and variety of examples that confused the issue. Acknowledging that even among the experts opinions varied as to what symbiosis meant, he redefined and classified its various manifestations.

For Schneider the only "true symbiosis" was an interaction between species at the physiological level. The intimacy and intensity of such a relationship would inevitably change the chemistry and even the physical makeup of one or both symbionts. Moreover, he saw that such a change must be controlled and passed on in a hereditary manner. Suddenly a new clarity of vision appeared. Schneider realized that symbiosis was far more than a curiosity in nature: it could create important evolutionary change.

In attempting to prove his theory, he faced the fact that symbiosis was more complex, and therefore harder to grasp, than Darwinism, a difficulty that would repeatedly hinder the understanding of this phenomenon. There was a wide range of symbioses. Although parasitism was an extreme form, it had to be included within the overall definition, even if this type of living interaction might lead to the destruction of one partner, the host. Mutualistic symbiosis, in which each symbiont possessed some property the other benefited from, was one of the most interesting forms, for it suggested that the survival potential of the interacting partnership might amount to more than the sum of the individual partners living alone. Take the lichens: the coming together of fungus and alga led to the creation of a composite life form that was better able to adapt and survive in a variety of environments than were the individual symbionts separately. Lichens are widespread in distribution from the tropics to the polar regions and are found in the lowest valleys as well as on the highest mountain peaks. In Schneider's view, symbiosis was best understood in the terms de Bary had first defined it: a living together of dissimilar life forms.

Schneider was certain that symbiosis could sometimes lead to changes in the structure and chemistry of the symbionts' bodies. Did such changes make possible the appearance of new tissues, new organs — even new forms of life? This question would be proposed, over the first two decades of the twentieth century, by a number of biologists, such as the Russians Andrei S. Famintsyn and Konstantine Merezhkovskii.

Famintsyn, the founder of Russia's first laboratory of plant physiology and a professor of botany at St. Petersburg University, became convinced, while investigating mutualistic symbioses between algae and other life forms, that he was observing major evolutionary change. What if the cells of every plant and animal on Earth had evolved from the symbiotic merging of smaller living organisms?

Over many years, Famintsyn struggled to investigate this question by extracting and attempting to grow chloroplasts, the organelles within plant cells that enable the leaves to capture the energy of sunlight. In a paper published in 1906 he claimed that although he had failed to grow chloroplasts, he had managed to extract and grow other organelles from the living cells of "lower animals." These, he thought, must have derived from free-living microorganisms that had made their homes inside the cells. He was now convinced that all of life had begun as "consortia" of simpler life forms.

Only a year earlier, Famintsyn's colleague and bitter rival, Merezhkovskii, working at the University of Kazan, had declared the symbiotic nature of chloroplasts. In 1910 Merezhkovskii coined the term "symbiogenesis" to signify evolutionary change as a result of symbiosis. Both Russians went on to perform a great many experiments, looking for hard evidence that the cell, and therefore all "higher" forms of life, derived from the symbiotic union of smaller life forms. Their experiments led to some limited successes, but these Russian symbiologists were no more successful in convincing their Darwinian-minded colleagues of their theories than were symbiologists elsewhere.

A few enlightened biologists scattered over the world plowed their obstinate furrows in this lonely field. For these few there was a truly inspiring consolation: growing evidence that symbiosis really did constitute an evolutionary force, quite different from that proposed by Darwin, even if the world at large ignored their findings. Indeed, after decades of resistance to all challenges, the world was also looking at Darwinism with a growing skepticism. There had always been doubts and criticisms within the mainstream of biology; and, after Darwin's death, in 1882, those questioning voices slowly gathered momentum.

As science moved into the opening years of the twentieth century, Darwinism was in crisis.

· 6 ·

DUELS AND GENES

> Nothing could be happier than this invention [survival of the fittest — Herbert Spencer's coinage] for . . . giving vogue to whatever it might be supposed to mean . . . It is the fittest of all phrases to survive.
>
> — GEORGE JOHN DOUGLAS CAMPBELL, "Organic Evolution Cross-examined" (1898)

IN AUGUST 1904, the biologist William Bateson arrived in London to attend a meeting of the British Association for the Advancement of Science. Bateson was to be nominated president at the meeting, but he had not arrived in a state of contentment. On the contrary, he strode in prepared for battle, with a six-foot-long chart full of controversy rolled up and slung, riflelike, over his shoulder. Among the more distinguished members of the audience were two colleagues with whom he shared an intense mutual loathing. One was Karl Pearson, the patrician mathematician, one of the founders of modern statistics and a supporter of Galton's radical eugenics. The other was Bateson's former friend and senior colleague at Cambridge University, Walter F. R. Weldon. The hall teemed with eager undergraduates, whispering among themselves in anticipation of a battle. The stage was set for the climax of what Michael R. Rose, professor of evolutionary biology at the University of California, Irvine, has called "one of the most destructive episodes in the history of biology."

People assume that great discoveries arise fully formed in the mind of genius, but in fact most evolve over time, with the discoverers having to wrestle with and modify their ideas long after they first put them forward. Darwin was no exception to this. Throughout his life, he struggled with various aspects of his evolutionary hypothesis, often changing his views on certain aspects and extensively rewriting his books to accommodate the changes. Moreover, although his theory is often portrayed as the single idea of natural selection, in fact it is a composite of related theories.

Ernst Mayr, regarded as the foremost living authority on Darwin, has subdivided Darwin's overall concept into five related theories: the general concept of evolution; the common descent of all life on Earth from a single ancestral life form; the diversity arising from the multiplication of species; gradual as opposed to sudden change; and the action of natural selection acting upon genetic variation. Those theories had differing fates when it came to acceptance by biologists.

Although Darwin himself considered all of these theories integral to one common theme, most evolutionists picked and chose among them. Even among those who believed in Darwin's evolutionary hypothesis, the term "Darwinism" took on a range of different meanings.

A major difficulty lay in explaining the differences between offspring of the same parents and between offspring and their parents. These differences, or "variations," were essential, according to Darwin's theory, if nature was to have enough variety to select from. And "variations" had to be inherited by future generations if they were to lead, over time, to new species. In Darwin's day nobody understood genetics, so he could only speculate that variation arose from a kind of "blending" of the pedigrees of the two parents. The first two chapters of *The Origin* are devoted to explaining how blending worked, both in animals and crops domesticated and bred by humans and in nature through the influence of natural selection. But over time, even Darwin became less convinced that blending was a sufficient explanation. In fact, in Mayr's words, "The origin of this variation puzzled him all of his life."

Another obstacle to general acceptance of his theory was Darwin's insistence on gradual change. As he explained it in *The Origin:*

> Why should all the parts and organs of many independent beings, each supposed to have been separately created for its proper place in nature, be so invariably linked together by graduated steps? Why should not Nature have taken a leap from structure to structure? On the theory of natural selection, we can clearly understand why she should not; for natural selection can act only by taking advantage of slight successive variations; she can never take a leap, but must advance by the shortest and slowest steps.

Before publication of *The Origin,* some of Darwin's stoutest defenders opposed his notions of gradualism. With what would appear great prescience to modern-day symbiologists, T. H. Huxley wrote to Darwin the day before the book was published: "You have loaded yourself with an unnecessary difficulty in adopting *Natura non fecit saltum* [Nature does not jump] so unreservedly." After *The Origin* was published, other supporters, including Francis Galton and the great Swiss embryologist Rudolf A. von Kölliker, urged him to modify his insistence on gradualism. In the words of Michael Ruse, "The appeal to large variations was certainly made more plausible because artificial selection working on small variations failed to produce specific changes, not to mention that no one had any direct evidence that natural selection in the wild changes one species into another."

Darwin, however, refused to budge. Like his critics, he believed that natural selection was only one of several forces driving evolution. But, impressed by the uniformitarian theories of the geologist Charles Lyell as much as by his own researches into human selection, he remained convinced that evolution depended on an incremental progression of small variations between individuals of the same species.

The diversity of life on Earth is staggering. More than 2 million species of plants have been named, and many more remain to be discovered. The total number of living species may be as high as 30 million, from the resilient chemosynthetic bacteria that take what they need for life from volcanoes miles under the ocean surface to the leviathans that for millions of years have swum over them. How could such diversity possibly have arisen from the relatively small amount of variation that might come from parental mixing? By the end of the nineteenth century, many scientists were openly skeptical of Darwin's theory.

Since the 1850s, biology had advanced a great distance, in terms of microscopic structure (histology), physiology, and biochemistry. Biologists were increasingly awed by the layer within layer of complexity in highly developed organs such as the eye and the brain. Even the simplest of creatures, such as the amoeba, turned out to be amazingly complex when they were studied at the level of microscopic and biochemical detail.

As knowledge of these fields grew, a number of eminent scientists were so awed by life's complexity that they doubted that any evolutionary theory was capable of explaining it. Skeptical creationists moved in for the kill.

One of these was the Swiss-American naturalist and geologist Jean Louis Rodolphe Agassiz, who had performed landmark work on glaciers and extinct fishes. Credited by some as the greatest teacher of biology of his day, Agassiz made no secret of the fact that he had little sympathy with Darwin's theory of evolution by slow and gradual changes selected by nature. Instead he taught his pupils that Darwin's theory was "a scientific mistake, untrue in its facts, unscientific in its methods, and mischievous in its tendency." Life, he claimed, came about through repeated acts of divine intervention.

Yet another attack came from studies of the fetus. Thomas Hunt Morgan was an American zoologist with a particular interest in embryology. From 1893 to 1910, he analyzed the patterns of the developing fetus with a view to understanding the governing mechanisms. He was a very knowledgeable scientist but, like most embryologists of his day, he found it difficult to explain how the remarkable embryonic adaptations could have evolved by the incremental addition of minute variations. At that time some evolutionists believed that the human fetus passed through all the stages of its previous evolution, a theory known as the "biogenetic law" and popularized by Darwin's German supporter Ernst Haeckel with the slogan "Ontogeny recapitulates phylogeny."

Scientists no longer believed that parental blending was the true mechanism of heredity. As an example, they pointed out that even if the blending of parental stocks gave rise to some kind of advantage in one offspring, the advantage would be diluted by half in every subsequent generation. Even more damning was the realization that blending could take place only through sexual reproduction. The earliest and longest period of evolution had involved simpler forms of life, such as the bacteria, which reproduce only by nonsexual means. The hereditary apparatus of bacteria did not undergo blending, so each offspring cell could only be identical to its mother. Yet bacteria also evolved. Nobody could offer a convincing alternative to blending, however, since genes and chromosomes were unknown and the central role of DNA was half a century into the future.

While heated debate was taking place in biological circles, August Weismann — a German biologist who would play a major role in the early understanding of genetics, in particular chromosome theory — became a passionate supporter of Darwin. He conducted a famous series of experiments on the markings of caterpillars, the shapes and colors of flowers, and the aquatic adaptations of marine mammals, using his findings to clarify and develop Darwin's theory. For twenty years he insisted that natural selection was sufficient to explain all of evolution. But leading evolutionists, including Huxley in England and von Kölliker in Germany, remained unconvinced, seeing Darwin's insistence on gradualism as an ideational blind spot. They believed that major jumps, resulting in changes all at once, must take place in evolution, a theory now referred to as "saltationism," from the Latin word *saltus,* which means a "jump."

That was exactly the argument that led to the confrontation in London in August 1904.

At the end of the nineteenth century, Francis Galton was the leading figure in the study of variation and a supporter of gradualism. His "Law of Ancestral Heredity" purported to show that an infinite potential for variety could arise from the blending of parental stock. Galton's thinking was taken up by Weldon, a close friend of Bateson's. After working happily together for years, however, Weldon and Bateson had a falling out. Half a century later, Bateson's assistant, R. C. Punnett (the geneticist who had lectured to the Cambridge branch of the Eugenics Society), recounted their story in a centenary address to the Genetical Society in Cambridge, outlining how it had all begun: "Both were convinced . . . that an attack must be made on the problem of [the origin of] species."

The problem between the two men was that Weldon, now a professor at Oxford, continued, like his friend Pearson, to accept Galton's theory of blending, while Bateson, more a naturalist than a statistician, became convinced that the variation necessary for evolution to work came from saltations. Pearson went on to develop statistical support for Weldon's gradualistic thinking, while Bateson took steps to confirm the saltationist perspective. Both were ambitious, combative men, and confrontation was inevitable.

With the gloves now off, Bateson gathered information from stockbreeders, bird fanciers, and horticulturists, and by 1894 he had accumulated enough material for a book, *Materials for the Study of Variation,* which was openly critical of Darwinian gradualism. Pearson and Weldon were incensed. Gradualism, based on the continuous variation afforded by blending, was so fundamental to their notion of Darwinism that they saw any criticism as a rejection of the whole concept of evolution. They fought Bateson with tongue and pen as viciously as Tennyson's bloody tooth and claw.

Four years earlier, while traveling by train to give a botanical lecture titled "Problems of Heredity," Bateson had happened to read some papers written thirty years before by an obscure Moravian monk named Gregor Mendel. Convinced by Mendel's reasoning, Bateson had rewritten his entire lecture.

Mendel's story is now well known. The abbot of an Augustinian monastery in Brünn, Moravia (now Czechoslovakia), he had a brilliantly logical mind, which led him, a farmer's son, to undertake highly original studies of the peas he cross-bred in the monastery vegetable garden. From these studies Mendel discovered the basis of what we now know as the laws of heredity. He found that certain characteristics of the peas were transmitted to the offspring in a predictable manner. These characteristics included tallness or dwarfishness, presence or absence of color in the blossoms or axils of the leaves, yellow or green color, and wrinkled or smooth skin in the peas. A single example will explain his line of reasoning.

When Mendel took the pollen from yellow peas and used it to fertilize the female parts of the flowers of green peas, the offspring peas were not a yellowish green, as one might have expected if parental characteristics blended. Instead they were all yellow. Even more intriguingly, when Mendel used the pollen from this new generation to fertilize the flowers of this same generation, the next generation of peas was no longer all yellow but a mixture of yellow and green, like the original parents. The ratio of yellow to green in that generation was not equal: there were three times as many yellow as green peas. By analyzing his results, Mendel showed that the inheritance of pea color could not be based on blending, as Darwin had believed; some *discrete* factors must be responsible for the two different colors. He had discovered what we now call genes.

Genes code for proteins, which become a part of the body's structure or play a part in its internal chemistry. They are the basic building blocks of heredity in much the same way that atoms are the basic physical units of everything in the world. We now know that some genes are "dominant," as was the gene that coded for the yellow color of Mendel's peas; some are "recessive," like the gene that coded for green color; and in some cases, both genes may express themselves, so the offspring exhibits an intermediary trait.

Mendel's findings, made as early as 1860, were of such fundamental importance that they still form the basis of our understanding of genetics today: we recognize that the genome, the genetic makeup, of any creature is composed of a discrete number of these building blocks. All that makes us human, as we have recently discovered, is a mere 40,000 or so genes, gathered together on 46 chromosomes.

Mendel reported his studies to the Brünn Society for the Study of Natural Science in February and March 1865, and his talks were published in the transactions of the society in 1866, just seven years after Darwin published *The Origin.* Although Mendel had read Darwin's book and had jotted down notes in the margins, Darwin was completely unaware of Mendel's experimental conclusions published in this obscure journal. Sadly, the truth is even more perverse than mere irony. Mendel corresponded with the distinguished Swiss botanist Karl Wilhelm von Nägeli and sent him his findings. But Nägeli, who shared with Pearson, Weldon, and Weismann an unshakeable faith in blended inheritance, did nothing to encourage the isolated Moravian monk, even misleading him by downplaying the importance of his work. Nägeli's prestige as a leading botanist guaranteed that nobody took much notice of Mendel's findings.

The unassuming monk was not alone in being cheated. In dismissing Mendel, Nägeli also cheated his hero Charles Darwin, who, had he but known of Mendel's experiments, would have found in them a much better basis for the hereditary aspects of his evolutionary theory.

On the train to his botanical lecture, William Bateson had realized the enormous importance of Mendel's experiments. Now, four years later, he felt ready to do battle in front of the crowded lecture hall at the meeting of the British Association for the Advancement of Science. Unfolding his six-foot chart, he talked about the evidence he had amassed in support of Mendel's discrete units of inheritance. He had repeated every one of Mendel's experiments with peas, confirming his results. He had found similar support for Mendel in experiments with mice. Where, he demanded to know, was the evidence of Pearson's Law of Ancestral Heredity?

The rest of the morning was given over to more experimental presentations.

Discussion moved on to inheritance in chickens. Bateson had slit open their unhatched eggs with his big, blunt-bladed knife, calling out to his wife, "Have you got that, Beatrice?" Loyal Beatrice would duly enter the details of down and comb, whether there was an extra toe or feathering on the leg, in the "Dead Book." Sweet peas were another hot topic of debate. Bateson grew thousands in his own garden, until his vegetable garden could no longer accommodate them, in spite of his wife's protestations that she needed her vegetables to keep the household alive. On then to the University Farm, where the experiments continued, flowers emasculated, their discrete hereditary characters crossed and recrossed. The results, read from the microscope perched on an old box from the farm, were entered in the book by Beatrice. All such experiments had further confirmed Mendel's concept of particulate inheritance.

At the meeting Bateson pressed home his advantage until everyone adjourned for lunch. When they returned, the room was packed full; even the windowsills had been requisitioned. The battle between Bateson and his opponents resumed, as Weldon, with sweat dripping from his face, stood up and addressed the audience in a loud voice.

Then, near the end, Pearson proposed a three-year truce between the two sides in the heated debate. The chairman, Reverend T. R. Stebbing, a mild and benevolent looking figure, stood to speak, deploring the ill feelings that had been aroused. The audience started to fidget, anticipating a tame conclusion to so spirited a meeting. But then he remarked, "You have all heard what Professor Pearson has suggested," adding, with a sudden swell of animation, "but what I say is let them fight it out."

The battle, which would rope in those most English of passions, the breeding of racehorses and even the gentlemanly art of cricket, was indeed far from over. Nine years later, in *Problems of Genetics,* Bateson, with extraordinary insight, examined how saltations might give rise to new species: "If we could conceive of an [infectious] organism . . . which may become actually incorporated with the system of its host, so as to form a constituent of its germ cell . . . we should have something analogous to the case of a species which acquired a new factor." This was very close to the proposals of symbiologists such as Merezhkovskii and Famintsyn with respect to the evolution of chloroplasts.

As so often with symbiotic explanations, nobody appeared to notice. By the first decade of the twentieth century, uncertainties about the mechanism of hereditary change had had such a negative effect that belief in Darwinism was at its nadir. By 1900 even Weismann was forced to agree with his long-time opponents that Darwinian natural selection based on the variation that arose from sexual mixing was not enough to explain all of evolution. In the words of Ernst Mayr, "The voices of Haeckel, Weismann, F. Müller, and Darwin's naturalist friends were merely cries in the wilderness, for the opposition to the mechanistic process of natural selection was almost universal."

What was needed, and needed urgently, to settle the argument in favor of a Darwinian explanation, was a new understanding of genetically based variation.

In 1900, the same year that Bateson had read Mendel's papers on the train, a Dutch biologist, Hugo de Vries, also had discovered Mendel's work. But de Vries thought of a remarkable application that Bateson had never considered. What if the discrete hereditary factors (what we call genes) could change? De Vries put forward a new mechanism of variation: the concept of random change in a unit of inheritance. Opportunity for change exists when genes are copied during reproduction, when a random change in the coding of a gene might arise from an error in copying, which de Vries called a "mutation." Logic would suggest that mutation was the natural ally of natural selection, providing Darwinian theory with the much-needed source of variation. But de Vries rejected natural selection and espoused an entirely different force of evolution. A convinced saltationist, he argued that mutations could produce a new species in a single leap, making natural selection redundant.

In time geneticists would confirm de Vries's theory of mutations, which occur at a low but fairly predictable rate. But today we know that a mutation affecting a single gene may have no effect at all, or it may decrease or increase the fitness of the individual in some way, usually to a very small degree. In time geneticists realized that most mutations are immaterial or even harmful. But occasionally one is "beneficial" in the evolutionary sense, and even though the benefit may be slight, a steady accumulation of such variations can cause significant change to a species. This realization took several decades to dawn; it wasn't until the 1930s, when a new mathematics of natural selection was put forward by R. A. Fisher and J.B.S. Haldane in Britain, by Sewall Wright and Theodosius Dobzhansky in the United States, and by S. S. Chetverikov in the Soviet Union, that the importance of de Vries's breakthrough was fully appreciated by mainstream biologists.

These scientists demonstrated that all that was required for selection of a mutated strain was that the mutation gave the offspring a 1 percent advantage for survival over others that did not possess it. Natural selection acting cumulatively on a steady series of these small variations could, over a long period of time, give rise to major evolutionary changes.

With this "synthesis" of Darwinism and Mendelism, Darwinian evolution reestablished its credibility. Today the majority of Darwinians perceive evolution as arising exclusively from the gradual accumulation of mutations and from sexual recombination, under the controlling influence of natural selection. This viewpoint, commonly known as the synthesis viewpoint or simply as "neo-Darwinism," has had an enormous impact on naturalists and experimental biologists, stimulating a new wave of evolutionary studies around the world. Indeed, for more than half a century, reductionist elaborations became the sole preoccupation of evolutionary biology, and the core philosophy of Darwinian competition was extrapolated to human behavior and psychology.

In symbiosis the mechanism of change is radically different from this Darwinian model. When two or more life forms interact, they bring together genomic and metabolic abilities that have already been honed by evolution. This interaction can involve a major evolutionary jump or saltation. Moreover, for Darwinians the mechanism of change (mutation) is essentially random and hence noncreative; while for a symbiologist, the mechanism of change is not random but a creative force in itself.

· 7 ·

THE INNER LANDSCAPE

In October 1838 . . . I happened to read for amusement "Malthus on Population," and being well prepared to appreciate the struggle for existence which everywhere goes on from long-continued observation of the habits of animals and plants, it at once struck me that under these circumstances favorable variations would tend to be preserved, and unfavorable ones to be destroyed. The result of this would be the formation of a new species. Here then I had at last got a theory by which to work.

— CHARLES DARWIN, *Autobiography* (1876)

IN THE LATE 1990s Jochen Brocks and Roger Buick, of the School of Geosciences at the University of Sydney, teamed up with Graham Logan and Roger Summons of the Australian Geological Survey Organization in Canberra, to search for fossils in the sun-scorched wastes of the remote Pilbara Craton in northwestern Australia. They were not searching for conventional fossils. Instead they were looking for "molecular fossils" that could be used to date the earliest evolution of complex life from preexisting bacteria.

If you examine the structure of a human cell, you find that it has the same basic plan as all complex life forms on Earth, including plants, animals, fungi, and even the unicellular organisms formerly known as protozoans: it has a nucleus, enclosed within a double membrane that contains the hereditary material — the genes packed together in chromosomes — surrounded by the fluid-filled "cytoplasm," where most of the routine chemistry of life takes place. This elemental organization is called a eukaryotic cell. In the opinion of Ernst Mayr, the evolutionary transition from the humble bacterium to the complex eukaryotic cell — a quasi-miracle of integration and coordination — was the single most important step in the history of life.

Evolutionists wanting to understand how this process happened posed many questions, two in particular. How early in the history of the planet did the first eukaryotic cells appear? And how, from such simple beginnings, was it possible for such exquisite complexity to arise?

The Australian scientists addressed the first of these questions. Taking great care to avoid contamination, they used a diamond drill to cut through 700 meters of rock until they found what they were looking for in shales that had been deposited during the Archaean period, an extremely ancient epoch of the Earth's evolution. They made an important discovery. In layers dating back an astonishing 2,700 billion years, they found molecular markers of membranes found only in eukaryotic cells. In the words of their report in the journal *Science,* "The biomarkers we report are the oldest known . . . They are more than a billion years older than those from the Barney Creek Formation, previously the oldest well-characterized molecular fossils."

These findings, if confirmed, would have great evolutionary significance. The appearance of the eukaryotic cell, even in primitive form, some 2.7 billion years ago implies far too rapid an evolution than would be possible through Darwinian gradualism alone.

In 1918, just five years after Bateson had imagined the evolutionary consequences of an infectious organism becoming incorporated into the germ cell of its host, a fifty-two-year-old French bacteriologist, Paul Portier, issued the most provocative challenge to evolutionary thinking since Darwin. In a book titled *Les Symbiotes,* he declared: "All living creatures, all animals from Amoeba to Man, all plants from Cryptogams to Dicotyledons, come about through the union of two different beings." He added: "Each living cell contains, within its protoplasm, formations which histologists call 'mitochondria.' To me these organelles are nothing other than symbiotic bacteria, which I call 'symbiotes.'"

Like the incredulous colleagues who first read these statements, we should take a mental step backward and consider the implications of Portier's claims.

He began with an observation made some thirty years earlier by the German physiologist W. Pfeffer. Only one life form on Earth, the bacterium, is capable of feeding independently: that is, it extracts life-giving energy from food that has not been processed by any other life form. Pfeffer called this independent nutrition "autotrophic." Portier saw in this fact an astonishing implication: since every other form of life is dependent in some way on food that has been processed by other life forms, then all of life must ultimately depend on the prior existence and continuing presence of these autotrophic bacteria.

In Portier's day, people took a very negative view of bacteria, which were known to cause tuberculosis, rheumatic fever, scarlet fever, puerperal fever, diphtheria, and bubonic plague. Publication of his book coincided with the end of World War I, when the war-torn populations of eastern Europe were ravaged by an epidemic of typhus, caused by a bacterium known as *Rickettsia* and spread by the body louse. This would eventually infect 30 million people and kill at least 3 million, while the pandemic of influenza, known as Spanish flu, carried off at least 20 million.

It was not altogether surprising that for the majority of scientists, as for the public, bacteria had only one role in life: a grim and evil role, as carriers of sickness and death. Any counterargument, that bacteria were often benign or even beneficial — let alone that they might be absolutely vital to life on Earth — seemed impossible.

Portier's book was received with incredulity and scorn. A year after its publication, his French colleague Auguste Lumière dismissed it derisively in *Le Mythe des Symbiotes.* Some of the criticism was warranted. For example, Portier claimed that mitochondria were the "ultimate units" of living matter. He set up experiments aimed at removing them from cells and he attempted to culture them in media, like any other bacteria. But Jan Sapp is convinced that Portier never went so far as to claim he could culture mitochondria outside the living cell, believing they had been adapted for life within the cell over eons of time. In Sapp's opinion, much of the subsequent criticism was misplaced, in part because Ivan Wallin, who could not read French, reconstructed some of Portier's statements as claiming that he had cultured mitochondria. Such was the unrelenting fury of microbiological orthodoxy that even thirty-seven years later, Paul Buchner, professor emeritus at the University of Munich and the greatest living German biologist, still felt the need to ridicule such notions, denigrating as "far-fetched" the hypothesis of those who believed the eukaryotic cell had evolved from the union of microbial forms: "Portier . . . began the controversy with his book *Les Symbiotes* (1918) . . . The wild flights of fancy he embarked on knew no bounds." Buchner was no scoffing Darwinian. The world leader in symbiology in his day, he pioneered studies of symbiosis in the lives of insects.

In spite of setbacks, new evolutionary theories continued to emerge. The German biologists Andreas Schimper and Richard Altmann had also proposed that chloroplasts and mitochondria were symbiotic lodgers inside living cells. A fellow German, Friedrich Meves, attributed all cellular differentiation to these structures. From Buchner's perspective, all "such 'errors' only served to retard our progress towards general understanding of symbiosis." Buchner went on to make abundantly clear what his own more modest understanding of symbiosis implied: "Independently of such extravagant concepts, in clear-headed unassuming work, the science of endosymbiosis has been laid stone upon stone . . . For us who have remained aloof from such speculations, endosymbiosis . . . represents a widespread, though always supplementary, device."

To shake the world, Portier and his German colleagues would have needed incontrovertible proof of the role of symbiosis in the origin of the eukaryotic cell. They just did not have it. And so the broad movement of science took no notice of such quaint notions as whole organisms physically merging.

In Portier's day there were serious conceptual obstacles to the acceptance of symbiosis as a major evolutionary force. One of the most formidable was that symbiosis extended into so many kinds of relationships, from the interdependence of different types of bacteria to the great cycles of life, such as the circulation of oxygen in the atmosphere. And the symbiologists were scattered throughout many disciplines, working in isolation in different countries and even continents, with little or no contact. Above all, they lacked a coherent voice. But slowly, over the decades leading to the 1950s, progress in other branches of science would make biologists stop and think afresh about the prejudices against bacteria.

By the 1930s, in agricultural colleges such as Rutgers in New Jersey, microbiologists were taking a different view. They knew that vast numbers of bacteria inhabit the soil, where they play a vital part in the carbon, sulfur, and nitrogen cycles. Ecologists realized that if the bacteria on Earth were to be wiped out tomorrow, all of life, including humanity, would quickly follow. Even disease-causing bacteria, such as staphylococci, live, for the most part, in a benign relationship on the skin or in the nostrils of their human hosts. The most lethal plague in history, tuberculosis, kills only a tiny fraction of the people it infects. Even today, in the opening years of the twenty-first century, tuberculosis exists as a latent infection in 1.7 billion people worldwide, yet it causes clinical disease in only 8 to 10 million and death in 3 million. It is necessary to remind ourselves that all such infections imply an interaction between parasite and prey: in other words, they are symbioses.

In 1974 the pathologist Lewis Thomas published his award-winning *The Lives of a Cell,* in which he supported the radical idea that infectious disease might be included within the umbrella of symbiosis. Thomas disagreed with the notion, still prevalent even then, that symbiotic organisms were merely enslaved. Symbiosis, to Thomas, was not the predatorial relationship predicted by the Darwinism of the nineteenth century, but something else entirely: a relationship that "seems especially equable."

In the popular series *The Science of Life*, the distinguished writers H. G. Wells, Julian Huxley (the grandson of T. H. Huxley), and G. P. Wells — all three convinced Darwinians — declared that many of the known symbiotic associations were actually driven by hostility. They examined a number of such symbiotic associations to show how they broke down under stress. As the situation changed, the nature of the interaction changed.

There was a good deal of truth in this statement. Mutualistic symbioses often evolve from parasitisms. Peter W. Price, professor of biology at Northern Arizona University, has even investigated and explained the selective mechanisms by which parasitism evolves into mutualistic symbiosis. Even after symbiosis is established, the interaction between the partners may sometimes involve a delicate balance to suit a certain environment, so that the balance changes if the ecological circumstances change. An example of this is found in the green hydra that lives in ponds and slow-moving rivers.

The hydra is green because of the presence of *Chlorella* algae within its cells. Angela Douglas, an evolutionary biologist at the University of York, in England, has made a careful study of this relationship, drawing attention to the fact that the hydra will survive even if the algae are bleached out by intense light. Comparing populations of green and bleached hydras under conditions of starvation shows that the symbiotic hydras survive in conditions where they have a good supply of light, whereas the bleached hydras die. But if there is a plentiful supply of food, the bleached hydras grow better. The explanation is that under starvation the algae provide the hydra with a sugar called maltose, which they manufacture using photosynthesis. But when the hydra has plenty of alternative food the maltose becomes unimportant, although the hydra still has to feed its algal partner. In other words, symbiosis may be useful much of the time, but under certain circumstances it can be a burden.

Even if this is true in certain cases, the overall concept remains. The modifications to an organism arising from natural selection are also derived from interaction with the environment: the modifications may fail or lead to further adaptation if the environment changes. For example, a lizard living in a desert climate will have adapted to a life of dry heat and scanty nourishment. If its desert ecology is subjected to a flood, this adaptation may no longer suffice, and those lizards that can swim best will be most likely to survive and to produce a new generation.

Any scientist who performs research in the field soon realizes that relationships are often messy and governed by multiple variables. Douglas sees the common denominator of symbiosis not as mutual benefit but as a novel metabolic capability acquired by one organism from its partner.

John Maynard Smith, emeritus professor of biology at the University of Sussex, and Eörs Szathmáry, of the Institute for Advanced Study in Budapest, both Darwinians, have noted, "It is a curious fact that one can kill aphids with antibiotics." They explain that aphids live by sucking plant sap, which lacks certain vitamins. The aphids obtain the needed vitamins from symbiotic bacteria that are transmitted from generation to generation of aphids by penetrating the eggs immediately after fertilization. This symbiosis, which has existed for at least 50 million years, is an example of mutualism: neither aphid nor bacterium can live without the help of the other.

For many life forms, symbioses with microbes help guarantee the supply of essential amino acids and vitamins. For example, legumes, such as peas and clover, form symbiotic unions with nitrogen-fixing bacteria, known as rhizobia, around their roots. This symbiosis plays a crucial role in one of the great cycles of life, that of nitrogen.

Nitrogen accounts for 80 percent of the Earth's atmosphere, but in this gaseous form it is not available for the myriad of metabolic processes that must incorporate elemental nitrogen. Nitrogen fixation is the essential step that makes the element available, yet this ability, widely distributed among the bacteria, is absent from all other organisms. Symbiosis between nitrogen-fixing bacteria and other forms of life offers an ideal solution. The bacterium gets the high energy it needs from its symbiotic host while the host gets nitrogen in a suitable form for its internal chemistry from the bacterium.

In 1927, nine years after Portier's contribution to symbiology, an American biologist took up the baton. Ivan E. Wallin, professor of anatomy at the University of Colorado School of Medicine, showed the crucial role of symbiosis in the biological structure and lives of many creatures, including corals, turbellarian worms, cockroaches, and even the Portuguese man-of-war. He also put forward new evidence, based on seven years of original research, that mitochondria really are symbiotic bacteria. Moreover, he asserted that bacteria are the true building blocks of life, "the primordial stuff" from which all higher organisms have evolved.

His arguments extended far beyond mitochondria. He proposed that symbiotic union between different organisms had led to the incorporation of chloroplasts into plants and that incorporation of motile bacteria had led to the locomotion of small organisms, such as *Euglena,* by means of the whiplike appendages known as cilia. He also suggested that the nucleus of the eukaryotic cell might have arisen from such a symbiotic union. Wallin went on to propose that the symbiotic inclusion of bacteria within cells might be a means by which new genes were added to the genome and that such genes might be transferred directly into the nucleus. This process would be a remarkable force for evolutionary change, radically different in nature from Darwinian theory. Indeed, Wallin believed that symbiosis and not Darwinian mutation was the explanation for the origin of species.

But it was difficult to explain certain aspects of how this type of evolution might work. For example, once the mitochondria or chloroplasts had been symbiotically incorporated into the cell, how were they transmitted to future offspring? This question — I shall call it "the enigma of heredity" — was a major hurdle to understanding and therefore credibility.

Wallin was well aware of the importance of such questions, but science in his day could not answer them convincingly. Like Portier before him, he encountered nothing but ridicule. Two of his colleagues declared that if mitochondria were in fact true living entities, it would "force a complete readjustment in our ideas of the organization of living matter and the applications to it of the laws of physics and chemistry." Wallin never wrote another paper on his evolutionary ideas.

In the same year that Wallin published his book on symbiosis, Hermann J. Muller published his first report on how the mutation frequency of certain genes in the fruit fly, *Drosophila,* could be increased 1,500 times, using heavy x-ray bombardment. Mutation became the fashionable research tool of genetics.

As we have seen, mutation was an extremely useful concept, giving Darwinians all they needed to revive their beleaguered theory. It provided a sound basis for the small increments of change necessary for nature to have a range of diversity to select from. In Muller's own words, the principle was reductively simple: "mutation, reproduction and the reproduction of mutation." And the finding that mutation itself could, in certain circumstances, be greatly accelerated gave an even firmer basis to natural selection.

Once again symbiosis was consigned to limbo.

· 8 ·

THE ENIGMA OF HEREDITY

> We are survival machines — robot vehicles blindly programmed to preserve the selfish molecules known as genes. This is a truth that still fills me with astonishment. Though I have known it for years, I never seem to get fully used to it.
>
> — RICHARD DAWKINS,
> preface to *The Selfish Gene* (1976)

THE PARAMECIUM is a single-celled life form almost as familiar to biology students as the humble amoeba. Familiarly known as the "slipper animalcule" because of its engaging shape, it can be seen as a grayish white speck moving slowly about in the water of ponds, where it feeds on the bacteria that digest dead leaves. It moves by means of rows of fine protoplasmic threads known as cilia, which lash at the water and enable it to swim and gather its food. In the 1940s this creature became a focus of great interest as a result of research conducted by Tracy Sonneborn, professor of genetics at Indiana University.

By that time it was becoming obvious that symbiosis existed in two broad forms: exosymbiosis (some call this ectosymbiosis), in which two or more species interact but remain separate, as in the partnerships of hermit crabs and anemones or the pollination of flowering plants by insects; and endosymbiosis, in which the interacting species actually combine physically, one, usually microscopic in size, entering the body of a second, usually much larger and referred to as the host.

From the point of view of genetics, endosymbiosis posed the greatest challenge and thus became the focus of a great deal of research into what came to be known as cytoplasmic inheritance.

Sonneborn realized that the cytoplasm must play an important part in heredity and the physical makeup of the individual organism. As Sapp says, "We have genes that make proteins, but how do the proteins organize themselves into supermolecular structures? Are we supposed to believe that genes make whole cells? Or to put it another way: if I give you all the genes and you shake them up in a test tube with all the other cellular components, do you think you are going to make a cell?" He is perfectly correct in pointing out that if all we have is genes, no matter how much we shake them up with all the other cellular components, we are not going to manufacture a living cell.

In fertilizing the ovum, the sperm inserts its DNA into an existing cell, the ovum, which includes not only the DNA of the maternal line but also the maternal cytoplasmic structures. When the fertilized egg begins to divide, the cytoplasm, with all of its component organelles and organization, faithfully reproduces with it. This cytoplasmic reproduction is a distinct event from what goes on in the nucleus. The cytoplasm of the ovum can be viewed as providing a living template of cellular structure in which the nucleus, with its DNA, resides. From the early days of theorizing about cytoplasmic inheritance, some evolutionists have wondered what this inheritance might really mean. In providing future offspring with a preexisting cytoplasm, with all of its component organelles, the ovum is contributing far more than its nuclear DNA. This cytoplasmic contribution was what interested Sonneborn. In his investigations of the genetics of the paramecium, he was setting out to contradict the most strongly held dogma of the world of biology: that the nucleus is the sole carrier of the genetic inheritance of the cell.

In one experiment, Sonneborn cut out a few rows of cilia and reimplanted them with the cilia facing in the wrong direction. Offspring paramecia, whether they arose through sexual or nonsexual reproduction, inherited this change of direction, and the change persisted through 200 generations. This kind of inheritance could not be attributed to nuclear genes. As Sapp explains, a "cortical" hereditary factor had been affected by outside influence, through mechanisms of cytoplasmic inheritance that are still largely unknown.

In another experiment Sonneborn studied the species *Paramecium aurelia.* Several strains of *P. aurelia* produce a poison that leaks out into the culture fluid, where it kills other sensitive species of paramecia. Sonneborn went on to mate killers with sensitives, a process that was possible only because the sensitives became temporarily immune to the poison during the act of mating. Mating and reproduction are often confused. In fact, they are not the same. Paramecia mate to blend nuclear genes, but they reproduce by the simpler process of binary fission. During mating, the parent generation shares its nuclear DNA, so each individual emerges with identical new nuclei while keeping its cytoplasmic structures exactly the same as before mating. If the killer and sensitive types were determined by nuclear genes, the parents should now be identical, as should all future generations. Sonneborn showed that not only did the parents maintain their traits of killer and sensitive after mating, but all subsequent generations inherited this difference: one mate still passed on its killer cells and the other its sensitive cells.

Another observation was even more revealing. Now and then two paramecia came together in a very different kind of mating, during which cytoplasm could be seen to cross from mate to mate across a connecting bridge. No nuclear material was involved in this. Yet after such conjugation, sensitive strains were changed into killers.

Clearly the hereditary factor that determined whether a paramecium was a killer or a sensitive was based not in the nucleus but in the cytoplasm. Sonneborn labeled this killer predisposition "kappa." Whereas nuclear genes would express themselves regardless of the environmental conditions in which the organism lived, variation in the environmental conditions seemed to influence the behavior of this kappa entity. Even stranger still, although kappa appeared to be determined by cytoplasmic inheritance, it influenced the whole organism. For example, it caused the paramecium's growth rate and even cellular divisions to be responsive to temperature and the available sources of nutrition. If the temperature was right and nutrition plentiful, killer paramecia ceased (temporarily) to be killers.

Another puzzling aspect was that the kappa entities appeared to reproduce independently of the reproduction of the organism as a whole — rather like mitochondria. Similar results were found in experiments with chloroplasts in *Euglena mesnili.*

It seemed that within the cytoplasm, and cohabiting with the organism, there were living entities that could be increased, decreased, or even eliminated entirely. Their elimination appeared to radically change the organism. To the Indiana geneticists, the conclusion was inescapable: some of the inheritance of the paramecium was stored and passed down to future generations through the cytoplasm. Sonneborn's experiments became the exemplars for others to follow.

At the base of every eukaryotic cilium and flagellum, the whiplike appendages associated with cellular locomotion, is a structure known as a "kinetosome." André Lwoff, at the Pasteur Institute in Paris, found evidence in single-celled creatures other than paramecia that the kinetosome could reproduce independently of any division by the cell as a whole. As with kappa in the paramecium, the environment influenced the kinetosome. In England, C. D. Darlington found the same capacity for self-duplication in centrioles, which play a vital role in mitosis. Self-replication was also found in mitochondria, chloroplasts, and many cytoplasmic bodies found in insects and plants.

But for Darwinians, convinced that the nucleus alone was the cradle of heredity, any such concept of cytoplasmic inheritance remained anathema. Sapp quotes the leading American biologist of the early twentieth century, Thomas Hunt Morgan, who aptly captured the prevailing view: "In a word the cytoplasm may be ignored genetically."

All such claims for cytoplasmic inheritance threatened evolutionary geneticists with the resurrection of what had once been the most formidable challenge to Darwinism, a rival they hoped was dead and buried.

In Paris, on December 18, 1829, an eighty-five-year-old naturalist died, blind and in penury, believing his work had been dismissed and forgotten by the uncaring world of science. Born of aristocratic parents, he had survived the terrors of the French Revolution to usher in the modern systematic approach to the study of nature. He had coined the term "biology" and, at the very least, had set in motion the conceptual revolution that was to follow. His name was Jean-Baptiste de Monet Lamarck and, for many French biologists even today, he, rather than Charles Darwin, was the true founder of evolutionary theory.

Darwin did not, of course, *invent* the concept of evolution. Varying notions of it had been around since the time of the ancient Greeks. But Lamarck popularized the concept some fifty years before Darwin.

Born in 1744, a time when belief in biblical creationism dominated Western culture, Lamarck intuitively grasped what we now call geological time. "Time," he wrote, "is insignificant and never a difficulty for Nature. It is always at her disposal and represents an unlimited power with which she accomplishes her greatest and smallest tasks." After years of studying plants and invertebrates, he put forward a theory of evolution in which life could be visualized as a series of staircases, from the simplest to the most complex. Humanity, in such a sublime and perhaps theologically inspired vision, was at the top of the uppermost staircase.

This same progression imbued his view of evolutionary mechanisms. Impelled by what he called "excitations" and "subtle and ever-moving fluids," Lamarck thought that the organs of animals became more complex. Perhaps there was a first glimmer of "adaptation" as a result of natural selection in his belief that organs became progressively strengthened and more complex with repeated use or weakened by disuse. Certainly he was mistaken in imagining that such changes were then inherited to become the basis of evolution. Lamarck believed that offspring could inherit a parental character developed through environmental interaction during the parent's own lifetime. The notion of hereditary change arising from repeated use was, for example, Lamarck's explanation for the long necks of giraffes, which derived from straining to reach high branches. In the same way, a blacksmith's son might inherit stronger muscles from the occupational exercise of his father.

Lamarck suffered for the revolutionary nature of his theories and was reduced to the humiliating status of scientific outcast long before he died. But interest in his theories grew after his death, and many leading biologists came to accept his thinking. Darwin, for example, advanced Lamarckian arguments in parts of *The Origin,* and Spencer was a firm believer in Lamarckian inheritance for most of his life.

But Lamarck had no more understanding of the mechanisms of genetics than Darwin, and his outmoded concept of the "inheritance of acquired characteristics" was abandoned by science when the synthesis of Darwinism and Mendelism placed the evolutionary emphasis on nuclear genes. From the orthodox perspective, it did not support the symbiologists' case that evolutionary change arising from the coming together of two different species during their lifetimes sometimes depended on the inheritance of acquired characteristics.

Throughout the 1950s and 1960s, the enigma of heredity became the center of a great deal of controversy. Meanwhile, for proponents of cytoplasmic inheritance it became increasingly clear that proving their theory would provide new information on how, perhaps as long ago as 2.7 billion years, symbiotic mechanisms might have played an important role in the greatest evolutionary transition of all, the origins of the nucleated cell.

As the inheritance controversy moved closer to the boil, it was caught up in a tsunami of change that was rushing through the world of genetics. The discipline was expanding to include leading-edge microbiology and, particularly, biochemistry. In England the discovery by Francis Crick and James Watson of the structure of DNA stimulated a new wave of nucleus-dominated genetic research, confirming and extending the gene-centered laws of Mendelian inheritance.

During this same period, fascinating information was coming from the relatively new science of virology. An earlier generation of microbiologists had regarded the viruses that infect bacteria as nothing more than parasites, taking over the bacterial genome to produce new viruses by infecting and then rupturing the cell. But as knowledge advanced, it became clear that while many viruses behaved in this way, others did not kill the bacterium they infected but entered into a long-term union of genomes, thus offering an extraordinary potential for evolutionary change.

For example, some viruses transform the bacterium's ability to survive certain stressful conditions, such as the arrival of an antibiotic in its neighborhood. Because viruses and other gene-swapping mechanisms could readily alter bacterial heredity, it was becoming difficult to rigidly classify bacteria according to their hitherto accepted taxonomies. Yet infection and heredity were still being viewed as disparate processes. Infection was merely a destructive force, the antithesis to life.

It was hardly surprising that the world of genetics was becoming confused. In 1952, in an effort to rationalize this confusion, the future Nobel laureate Joshua Lederberg published an overview explaining why the nucleus could no longer be regarded as sacrosanct and why geneticists should consider a new idea of "infective heredity," which included the potential of viruses to change the genetic pool of infected bacteria. In this paper Lederberg proposed the term "plasmid" to embrace all extrachromosomal genetic elements. A plasmid might be a virus or any of a number of other genetic packages comprising whole genes or parts of genes. Lederberg went on to show how such genetic elements could move from cell to cell — or from one organism to another. When plasmids entered the cytoplasm, their actions and subsequent evolution were symbiotic. At that time the great population geneticist Theodosius Dobzhansky was also considering the possibility that all of the hereditary components found in the cytoplasm were symbionts.

The evolutionary origins of mitochondria now moved center stage. Cytoplasmic inheritance had an obvious appeal to those promoting the idea that mitochondria had evolved through symbiotic incorporation of bacteria. For example, in endosymbiosis, in which one symbiont lives in the cytoplasm of its partner, the cytoplasmic symbiont could reproduce by binary fission, providing an independent mechanism of inheritance that could exist side by side with nuclear reproduction. Indeed, confirmatory evidence of this scenario was rapidly accumulating, with new research suggesting that mitochondria had retained at least part of their former bacterial genomes.

Genes in the cytoplasm! The Darwinian reaction was to dismiss this idea as nonsense, but the evidence in favor of it continued to grow. The Darwinians then argued that if genes were associated with these cytoplasmic inclusions, the only plausible explanation was that they had been borrowed from the nucleus. In 1950, when the Genetics Society of America held its golden jubilee at Ohio State University, the *Drosophila* expert Hermann Muller took the view, then prevalent among Darwinians, that if mitochondria and chloroplasts had genes, these were just genetic debris from some early evolutionary stages of the organism, before the primitive cell had sequestrated its genetic material in the nucleus.

Like most biologists of his day, Muller wanted to keep things simple: an autocratic nucleus, sacrosanct within its membrane, governed all of the genetic apparatus of the cell and controlled all the biological processes taking place in the cytoplasm.

But the truth was not so simple and elegant. The interdependence of organisms in symbiotic associations was so vague and wide-ranging that it blurred the boundaries of taxonomic definition: where did the individual organism begin and end if genetic material could arrive from beyond the cell walls and change an organism's heredity? But now Lederberg was reinterpreting Portier, declaring that all heterotrophs — in other words, all of life other than certain bacteria and green plants that could extract their nutrients from the non-organic environment — were "genetically insufficient." It was possible to construct a graded series of symbioses as genetic interactions, from the genes inhabiting a single chromosome, to plasmids, and even to extracellular (and extraorganismic) ecological associations with varying degrees of stability and specificity. Biological science would have to reappraise every great transition along the march of evolution, from the earliest beginnings of self-replicating entities to that most remarkable of sentient primates, humanity itself.

When Lederberg raised such possibilities, the scientific world was forced to take some notice. A formidable figure in microbiology and genetic thinking, he could not be dismissed in the way so many earlier symbiotic thinkers had been. But even Lederberg did not wish to proclaim this as anything beyond a possibility; as yet, nothing was certain.

As late as 1972, two biologists at the University of Indiana, R. A. Raff and H. R. Mahler, refuted all evidence for cytoplasmic inheritance in a paper titled "The Non-Symbiotic Origin of Mitochondria." These authors claimed that mitochondria had never been bacteria. Rather, they had evolved from specialized invaginations of the cell membrane in the distant past, acquiring some genes from the nucleus through the mediation of a plasmid. Three years later, in a manner reminiscent of the dismissal of Portier by Lumière, they poked fun at the concept of mitochondria arising from bacteria in a book chapter entitled "The Symbiont That Never Was: An Enquiry into the Evolutionary Origin of Mitochondria."

Even at this stage, in the words of Jan Sapp, "To call a particle a symbiont amounted to name calling." What was now needed was a new, clear perception of the role of symbiosis that made sense in an expanding era of molecular biology. And the person who provided this leadership was Lynn Margulis, who would later become professor in the department of geosciences at the University of Massachusetts at Amherst. Although at this time she was no more than a graduate student, she was also a symbiotic thinker of great innovation. Equally important, she was a woman of courage and tenacity, prepared to stand up and fight from her corner against a skeptical and, from a symbiotic perspective, largely ignorant neo-Darwinian world.

· 9 ·

SYMBIOSIS COMES OF AGE

> The writings of contemporary symbiosis researchers are replete with episodes and anecdotes about how their intellectual ancestors were ignored and ridiculed by an unappreciative scientific community.
>
> —JAN SAPP,
> "Evolution by Association: A History of Symbiosis"

ONE DAY in 1991 Lynn Margulis presented a lecture to the American Association for the Advancement of Science. Her audience ranged from students struggling to come to grips with life's taxonomy to combative fellow leaders in the world of biology. Most biologists illustrated their lectures with attractive animals, birds, or mollusks, but the slides Margulis projected onto the screen were of the microscopic creatures called protoctists — a vast kingdom of nucleated microorganisms, including those formerly known as protozoans. These organisms had been the focus of her research for decades. She had long bemoaned the fact that the scientific world had so ignored this entire kingdom that only a few had names (amoeba, euglena, paramecium, spirogyra). In the words of John Maynard Smith, "She knows an incredible amount about strange beasts most of us don't know anything about."

Margulis began by warning her audience that the conventional systems of biological classification concentrated too much on the larger animals, creatures that rightfully should be classified as only a small branch of the subdomain of the eukaryotes. In part this incorrect emphasis was the reason evolutionary biology had become bogged down in dependency on neo-Darwinism. She went on to illustrate her point by explaining the diversity of microbial life, including predation, photosynthesis, communication, social organization, and motion. Suddenly, in discussing the evolutionary origins of eukaryotic life from bacterial precursors, she became openly combative.

"Neo-Darwinism," she declared, "is misleading. I see no evidence whatsoever that these changes can occur through the accumulation of gradual mutations."

Of course, many in her audience were neo-Darwinians. Exasperated by their silent skepticism, she issued a challenge: Would anyone care to name an unambiguous example of a species that had been shown to evolve by the building up of chance mutations? One man stood up to name a species of corn, only to be contradicted by another. Where, then, she asked, almost a century and a half after Darwin first put forward his theory, was the overwhelming evidence that gradual change brought about by the accumulation of chance mutations is responsible for the origin of species?

"See for yourself!" she proclaimed, projecting a slide of a "red tide" microorganism that lives on the surface of Finnish lakes. Visible in the cytoplasm of each translucent body were tiny bodies, the vestigial remains of a smaller microorganism called a cryptomonad. "Long ago," she explained, "one of these guys ate but did not digest the other. Now they require each other to reproduce."

She added, "I can give you a dozen of these examples — and you give me a type of corn . . . maybe. You give me maybe — I give you the evidence. So why do you think that you are right and I am wrong?"

In science, just as any other field of human endeavor, personalities become important.

Lynn Margulis first met the late astronomer Carl Sagan when she was fourteen and attending an undergraduate course at the University of Chicago. He was a graduate student with a physics degree and she was studying for her A.B. in liberal arts. She was charmed by Sagan's single-mindedness. "He just taught us that we chickens could contribute to the scientific world, the scientific enterprise." They married and later divorced. By the time she was pregnant with their son Dorion, Margulis had switched to biology, in particular evolutionary genetics. Unimpressed by the "overly abstract neo-Darwinian concepts" then promulgated almost as religious dogma in population genetics, she turned to genetic systems that other scientists were inclined to ignore: "The data on cytoplasmic genes fascinated me from the time I first learned about them." Inevitably this interest carried her toward symbiosis and the controversy that surrounded it.

At college her instructors adopted a revolutionary attitude to teaching. Eschewing textbooks, they insisted that the pupils read scientists' original papers, so they could discover for themselves the actual thinking of Newton, Galton, and Mendel. This single-minded approach would become important in Margulis's future career. In 1957 she moved to the University of Wisconsin, where she studied genetics with James Crow, "the best teacher in the world." Joshua Lederberg, who taught in the genetics department, later asked her why she fell asleep in morning lectures. She replied, "When I go home, I am nursing children."

At Wisconsin Margulis became aware of the inconsistencies in the prevailing ideas about heredity. Green female plants crossed with white males of the same species sometimes gave rise only to green offspring. Yet if green males were crossed with white females, the plants that grew from the fertilized seeds were all white. If inheritance arose solely from nuclear genes, with random mixing during sexual fertilization, it would not matter which parent contributed which color trait. It was becoming ever more obvious that cytoplasmic factors, presumably genes, must play an important role in the germ cells of animals and plants.

Mitochondria are vital components of nucleated cells, where they enable the use of oxygen in respiration. And it was now known that the mitochondria of animal cells contained genes. Genes had also been found in the chloroplasts of plant cells. Without question, both mitochondria and chloroplasts were passed down through generations as part of reproduction. Yet they came from only a single parent, usually the female, and the inheritance was not through the nucleus.

For her master's degree, Margulis cut up amoebae to determine if any DNA or RNA activity remained in isolated portions of cytoplasm. She soon had her answer: RNA played an active part in the cytoplasm, and DNA was also present. Later this DNA would turn out to mark the presence of bacterial bodies without walls. During an interview she remarked to me, "That was my first experience of cytoplasmic inheritance." In 1960, after completing her master's degree at Wisconsin, aged twenty-two and already the mother of two young boys, Margulis enrolled as a graduate student in the genetics department of the University of California at Berkeley.

By now she was intrigued by the abundance of symbiotic interactions in nature, particularly those involving bacteria living with — sometimes inside the cells of — insects and worms. Like others before her, she suspected that the mitochondria and chloroplasts were remnants of what had once been free-living bacteria. "It seemed obvious to me that there were double inheritance systems with cells inside cells." She became fascinated by Tracy Sonneborn's work on cytoplasmic inheritance in paramecia, grasping at once that his experiments had confirmed that characteristics acquired by the organism in its lifetime could be passed on: "They confirmed a kind of neo-Lamarckian inheritance."

Margulis searched for and found many other examples of cytoplasmic inheritance in which the agents of inheritance were not "naked genes" left over from some past evolutionary accident but elements within the vestigial bodies of once free-living bacteria, now incorporated as organelles within the cells. Living survivors of a more ancient eukaryotic cell, one of the archaeprotists, had a nucleus containing rod-shaped chromosomes, but with cytoplasm that contained no mitochondria. It was not far-fetched to wonder if some archezoan had once swallowed up a bacterium, which then entered a symbiotic partnership with the cell, remaining forevermore within its cytoplasm as the mitochondrion.

The familiar mitochondria and chloroplasts still resembled bacteria in their behavior and metabolism. In fact there seemed little difference between a bacterium newly trapped within a cell and a mitochondrion inherited as part of cellular evolution. What everybody called a chloroplast was simply a blue-green bacterium, known as a cyanobacterium, that had shed its cell wall to reside inside the cytoplasm of a plant cell.

Emboldened by her survey of the literature, Margulis predicted that if mitochondria and chloroplasts had once been free-living bacteria, they might still retain some of their bacterial DNA.

Bacterial DNA is identifiable in its structure and genetic coding. If her prediction proved to be correct, then the origin of these cytoplasmic bodies could be absolutely confirmed. Her deduction was subsequently proven by geneticists around the world. Mitochondria not only reproduce independently of the nucleus, by binary fission, they also have their own DNA, which, as in bacteria, takes the form of a single circular molecule. They also have their own messenger RNA and transfer RNA, which form part of the protein-manufacturing factories called ribosomes. Today scientists accept the evidence that mitochondria and chloroplasts were once free-living bacteria, related to surviving forms on Earth. Some chloroplasts can be distinguished from existing cyanobacteria only with such difficulty that the distinction between them is biologically meaningless.

With this new understanding, the concept of symbiosis came of age. Mutualistic symbiosis, for example, no longer implied some cuddly living together but a metabolic interaction or dependency between different kinds of organisms, on which their survival depends and from which great evolutionary changes may arise.

If current estimates are correct, the hard bargain that resulted in the evolution of mitochondria took place somewhere around 2 billion years ago. That original endosymbiotic union was similar to what Kwang Jeon witnessed in his study of the amoeba and the X-bacterium. The particular bacterium involved is still uncertain, but recent gene mapping of bacteria has suggested an extraordinary candidate.

Epidemic typhus fever is one of the most notorious plagues of history. Commonly associated with famine and civil unrest in the wake of war, it caused the major epidemic already referred to in the wake of World War I in Europe. The causative germ, one of the rickettsias, is passed from person to person through the bite of the human body louse. This rickettsia is extremely unusual. Halfway between a bacterium and virus in size, it resembles a virus in that its life history takes place inside the infected host cell, a so-called obligate intracellular parasite. In evolving its endosymbiotic lifestyle, the germ has lost a good deal of its metabolism, along with the coding genes.

Michael W. Gray, professor of biochemistry and cellular biology at Dalhousie University in Halifax, Nova Scotia, and one of the foremost experts on the evolution of cytoplasmic organelles, believes that the typhus-causing germ is the closest living relative that we know of to our human mitochondria. This does not necessarily mean that our mitochondrial ancestor was the typhus-causing rickettsia; in time, even closer relatives may be found.

Some important deductions about the story of evolution can now be drawn. One of the most intriguing is that all eukaryotic life, including many of the 250,000 species of protoctists, the fungi, and every plant and animal species on Earth, must have evolved from that singular endosymbiotic merger of two types of bacteria. If leading authorities on bacterial evolution are correct, other commonalities of origin may go back to the very beginnings of life. Scientists have found similarities in the genetic sequences of a highly conserved form of RNA in hundreds of early life forms, suggesting that the most ancient bacteria, the modern bacteria, and the eukaryotic cell all have a common origin.

In 1961 the evolution of the nucleated cell was one of the greatest enigmas in biology. So, only one year into her Ph.D. studies, Margulis became very interested in the possibility that symbioses of this nature might provide the answer.

At Berkeley, she was surprised to find that the geneticists, with a single exception, had no interest in evolution. But Curt Stern assigned her to read up on the genetics of the alga *Chlamydomonas*, which contained mitochondria and chloroplasts within its cytoplasm. "It was a gorgeous organism from the point of view of evolutionary logic. All I wanted to do for my doctorate was get the DNA from the chloroplast to prove the basis of cytoplasmic inheritance. But when I took this idea to various people, they told me, 'You're looking for Father Christmas!'" At this stage she became very interested in Sonneborn's studies of the paramecium.

By the early 1960s, evidence for the symbiotic origins of mitochondria and chloroplasts was steadily accumulating. Mitochondria were killed off by streptomycin, a drug used to treat bacterial diseases such as tuberculosis. Chloroplasts very closely resembled the common photosynthetic cyanobacteria. The nucleus was being intensively explored in every major genetic laboratory. But the "cortical" inheritance discovered by Sonneborn remained a mystery. One day Margulis had her "Eureka" moment. As she would recall later, "I remember, as a graduate student, I was standing in the library at Berkeley, poring over the Sonneborn papers and the literature on chloroplasts, when I came across a review of a book by Herbert F. Copeland." Copeland, a botanist, taught biology in Sacramento. In his book *Classification of the Lower Organisms,* published in 1956, he suggested a new way of looking at the tree of life. One reviewer wrote, "This classification is so idiosyncratic that this reviewer has great difficulty in evaluating it." In time this would encourage Margulis to consider that a classification system, or taxonomy, might be useful in her attempts to solve the evolutionary enigma.

Margulis had to go to the Library of Congress in Washington to read Copeland's book. As big as a telephone directory and dense with detail, the book was based on Copeland's studies over thirty years. One organism he discussed was a protist (a single-celled protoctist) symbiont that lives in the guts of termites. The symbiont surface was so covered with wriggly, corkscrew-like bacteria, called spirochetes, that the bacteria had been mistaken for cilia and the protist mistakenly classed with the ciliates. As Margulis recalled to me, "When I first saw that, I said to myself, 'That's the last symbiont I need. That's the ciliate symbiont.'" She was now sure she had the information needed to advance a theory of how a series of bacterial mergers gave rise to the first nucleated cells.

Many of the single-celled life forms known as protists still live in an atmosphere without oxygen, for example in the intestines of plant-eating insects and herbivores. This anaerobic existence suggested to Margulis that anaerobic protists represented the first stage in cellular evolution, the stage before the incorporation of mitochondria. To explain how the first anaerobic protist came into being, Margulis proposes an initial union between a freely mobile modern bacterium (known as a eubacterium) and an ancient bacterium (an archaebacterium) that could breathe sulfur rather than oxygen and also tolerate extreme heat.

If Jochen Brocks and his colleagues were right in their interpretation of the molecular fossils they found, this union, forming the primal substance of all nucleated cells, must have taken place more than 2.7 billion years ago. Then the two bacterial genomes became incorporated in some way within a separate entity from the rest of the cell, enclosed within a membrane, perhaps even forming, or helping to form, the first nucleus. The "modern" bacterium was able to swim and was probably one of the corkscrew-shaped spirochetes mentioned in Copeland's book, a eubacterium that could move through water by wriggling its body. Its ancient partner would have been similar to one called *Thermoplasma,* an archaebacterium that is found today in the hot springs of Yellowstone Park, which may well have breathed sulfur rather than oxygen. *Thermoplasma* supplied the spirochete with its hydrogen sulfide wastes, and the spirochete supplied the consortium with the capability of rapid movement. If Margulis is right, the spirochete's attachment sites later evolved into cellular structures known as centrioles. These play a vital role in one of the most beautiful mechanisms in biology, the complex ballets of cellular reproduction known as mitosis and meiosis, in which the centrioles help orchestrate the "dance of the chromosomes."

Initially, the sulfur-loving, heat-resistant, freely mobile life form could not breathe oxygen. With the evolution and global spread of the cyanobacteria, which generate oxygen, that element became a significant component of the Earth's atmosphere, encouraging the evolution of oxygen-breathing bacteria. One of these oxygen-breathers became the mitochondrion, through an endosymbiotic union with the previously anaerobic protist. This larger, more complex protist acquired the ability to take in particulate food, much as an amoeba does today. So, as Margulis explained, a new "complex and startling being" began to spread over the shorelines of the Earth about 2 billion years ago. Some of these protists entered into an endosymbiotic union with cyanobacteria, resulting in the first aquatic green algae, the forerunners of all the plants.

Natural selection acting on gene mutations cannot create new genes; it can only modify those that already exist. These formative symbiotic unions, on the other hand, involved the merging of thousands of genes, every one of which had already evolved over an eon or more, into a new hybrid organism. Although that scenario is not quite what Lamarck had in mind, Margulis sees it as an example of neo-Lamarckian evolution that is orders of magnitude greater in its potential for change than Darwinian gradualism. This process of change is so sudden and spectacular that she has compared it to an evolutionary lightning strike.

There is another implication of Margulis's theory for evolutionary biology. Neo-Darwinians believe that evolution takes no definite path toward increasing complexity. Symbiologists take a very different perspective: with each of these endosymbiotic steps, the resulting hybrid involved a huge increase in genetic and biological complexity.

After years of orthodox rejection, Margulis's first paper describing this series of endosymbiont unions was published in 1966, and it became the basis for her now famous "serial endosymbiosis theory," or SET. Hailed by Sapp as the most daring and serious effort in pursuit of the symbiont hypothesis during the 1960s, SET proposes that the nucleated cells of plants and animals, as well as those of fungi and the protoctists, originated through mergers of different types of bacteria and that the hybrid forms evolved in a specific sequence of symbiotic steps. Given the broadly hostile reaction to symbiosis over the previous century, it will come as no surprise that Margulis encountered a wall of skepticism. But in time, as she refined and extended the concept, even the most doubting Darwinians not only accepted her theory, they marveled at the beauty of it. As Richard Dawkins generously remarked, in *River Out of Eden,* the serial endosymbiosis theory of the origin of the eukaryotic cell is "incomparably more inspiring, exciting and uplifting than the story of the Garden of Eden . . . Like most biologists, I now assume the truth of the Margulis theory."

But we should not get too carried away by this story of the triumph of symbiosis. The eukaryotic cells in the fingers that hold open these pages, or in the cells in the retinas that scan them, are more exquisitely wonderful than mere unions of bacteria: they are labyrinthinely complex miniature worlds, and like worlds they incorporate all manner of evolutionary mechanisms. In their influential book *The Origins of Life,* Smith and Szathmáry put forward a composite of Darwinian and symbiotic mechanisms that enlarge on SET; they advance reasonable theories of how the nucleus, and even the physical construct and concept of the cell itself, could have arisen through essentially Darwinian mechanisms.

Common sense would suggest that evolution has embraced both neo-Lamarckian saltations in the form of symbiosis and neo-Darwinian gradualism through the accretion of useful mutations.

Certain aspects of SET remain controversial. For example, Hyman Hartman, working at the NASA Ames Research Center at Moffett Field, California, has suggested that the nucleus began as a free-living organism. And there are extraordinary new discoveries that suggest that the role of viruses in many aspects of cellular evolution has been greatly underestimated. Such is the nature of science that hypotheses are put forward to be proved or disproved, often resulting in further honing and modification of the theory. As a scientist, Lynn Margulis does not pretend to be above criticism or even refutation: "I am prepared to be incorrect."

But if she is correct in her spirochete hypothesis, there may be implications for the evolution of the human brain. Our nerve cells, like those of all animals, have threadlike extensions called axons and dendrites, which are based on a microtubular infrastructure that has similarities to the spirochetes that fused with the Archaebacteria in SET. If these did originate with the incorporation of a spirochete a billion or more years ago, then we might even owe the evolution of our human brains to that ancient symbiosis.

· 10 ·

THE WONDER OF SYMBIOSIS

> To look still more broadly, we discovered that terrestrial life is a dense web of genetic interactions.
>
> —Joshua Lederberg

Coral reefs have been called the rain forests of the sea. In these precious domains, swimmers equipped with no more than masks and snorkels can watch creatures with the luminescent colors of gems engage in behavior rituals as enchanting as the mating of the most splendid birds.

The builders of these reefs are small marine animals that secrete their own external skeletons; the slow accumulation of these skeletons over millions of years forms the framework of the reefs. Members of the phylum Cnidaria, which also includes the jellyfishes and anemones, corals exist in a multitude of forms. All are colonial animals equipped with stinging tentacles (*cnida* is the Greek word for stinging nettle) that emerge like opening flowers to catch tiny prey and drag it into the saclike stomach for digestion. On the 2,000 square kilometers of the Great Barrier Reef of Australia, there are about 350 coral species, including both hard and soft varieties. These reef-building corals grow in clusters, extending in a confluent surface layer over the graveyards of past generations. Each species has a different shape or color or blend of colors; in association with other related cnidarians, they create all manner of sculptural forms. Some fashion huge rounded domes elaborated with sulci, like the human brain; others form twigs, bifurcating into staghorns or wide fan-shaped arborescent branches or the most exquisitely floral forms, all decorated with every subtlety of tone and hue.

Of course, coral reefs are much more than just a beautiful curiosity. They are the homes and nurseries for almost a million species of fish and other marine creatures and algae, many of which we rely on for food.

Nature films about the coral reef ecology used to focus on the predatory activities of its many residents, and certainly a reef can be a dangerous territory to inhabit. But lately the focus has become more balanced as filmmakers cater to a public delighted by the many symbiotic relationships among its life forms. Reef-building corals are exclusively symbiotic with the unicellular yellow-brown photosynthetic organism known as dinomastigotes, which live their entire lives inside the cells lining the coral's gut. The coral's stinging tentacles are predators by night and the dinomastigotes capture sunlight by day, storing its energy in the form of carbohydrates. The ecosystem depends on this symbiosis, for the dinomastigotes supply half or more of the coral's energy needs, vastly increasing its ability to lay down calcification. This symbiotic union is believed to have evolved in the late Triassic period, some 210 million years ago.

Another group of tiny aquatic life forms is known collectively as the Foraminifera. With over 35,000 described species, "forams" are the most widespread of the organisms in the oceans. The individual shells are no bigger than grains of sand, yet in their billions they form an important component of plankton, the basic food source of all marine life. They also incorporate intracellular symbioses with a wide variety of photosynthetic microbes, which make the foram bigger and strengthen its shell, thereby improving its ability to survive. The symbiotic union is so self-sufficient it needs only a few additional vitamins to survive and reproduce. Mutualistic symbioses like coral and forams are important components of life in the sea.

On land, one of the five kingdoms of life, the fungi, comprise some of the most ancient terrestrial life forms. Like bacteria, fungi are essential to the cycles of life. Underneath the soil certain fungi form immense labyrinths of filamentous mycelia, which can extend for miles and which play vital roles in plant nutrition. Aboveground these fungi emerge as the fruiting bodies we recognize as mushrooms, which cast vast quantities of spores into the air. So enormous are the numbers of fungal spores and so efficient is their distribution they are found at every level and within every crevice of the biosphere. Fungi cannot live without carbon, which must be obtained from other creatures, whether bacteria, plants, or animals. This is why fungi are found growing in and around other forms of life, on their secretions, excretions, dead flesh, and — in saprophytic, parasitic, or mutualistic symbioses — in intimate relationships with a vast array of life forms.

As we have seen, fungi are an integral part of lichens. Even more important, most land plants have evolved from a joint venture between fungi and green algae, a relationship that evolved over hundreds of millions of years to become the mycorrhizae that nourish the roots of most species of plants today. Only recently have scientists realized that a single pine tree may have several different fungi in a mutualistic symbiosis with its roots. Peter R. Atsatt, a professor in the department of ecology and evolutionary biology at the University of California, has proposed an even more radical concept: that land plants may have arisen from the very early incorporation of a fungal genome into that of a green alga and that this hybrid organism went on to play a central role in the evolution of embryos, then of pollen and seeds.

Many fungi enter into complex symbiotic relationships with insects, in which the fungus helps break down the tough cell walls of plants so that the insect can digest them, while in return the fungus has a seat at the insect's dining table. Among the many species of insects that have mutualistic symbioses with fungi are scale insects, gall midges, wood wasps, and anobiid beetles.

One delightful example of such a relationship concerns the *Atta* ants of Central and South America, which harvest leaves and carry them back into their subterranean nests. After the ants masticate the leaves, they are further digested by a domesticated fungus growing in the nest's inner garden. The fungus breaks down the leaves' cellulose walls, liberating their hidden stores of nourishment for the ants, while the ants give the fungus protection and shelter and a constant supply of leaves to feed on.

Almost a third of all known species of fungi are involved in mutualistic symbiosis of one form or another, which clearly has evolutionary implications. In the words of Bryce Kendrick, "Several of these relationships have given rise to major evolutionary innovations, which have conferred on the interdependent organisms the ability to colonize habitats previously unavailable to them." In this way, symbioses of many different types made possible the expansion of life into hitherto barren and hostile ecologies in the water and on the land surfaces of the primal Earth.

Phyla are the major divisions, below kingdoms, in the hierarchical classification of life. All members of a phylum share a fundamental physical or physiological characteristic, whose evolution gave rise to that phylum. In 1987 Lynn Margulis and her colleague David Bermudes suggested that many, perhaps all, of the fundamental characteristics that define each phylum came about through the creativity of endosymbiosis.

They showed how twenty-eight of the seventy-five phyla, excluding bacteria, depended on the incorporation of organelles that had once been free-living microbes. These include thorny-headed worms, the Cnidaria, and the exotic creatures known as Ventimentifera, which live around the deep sea vents; they also include all mycorrhizal plants, including conifers, angiosperms, cycads, and ginkgos, as well as fifteen protoctist and two fungal phyla. Lower down the taxonomic system, symbiotic mergers have given rise to legumes with their nitrogen-fixing rhizobial bacteria, many wood-eating cockroaches and termites, as well as all ruminant mammals and the luminous fish that inhabit the ocean deeps.

In all these cases the enigma of heredity demanded an explanation that was far more varied and complex than Darwinian mutation alone. Each case of symbiosis had to be examined in its own light, a daunting prospect for evolutionary geneticists. In some cases it was clear that the symbiont was passed down through bacterial-style binary fission within the cytoplasm. Richard Law, a lecturer in the biology department at the University of York, described the hydroid *Myrionemia amboinense,* whose eggs carry algal cells within them. Giant clams have a more complex relationship with the dinoflagellate *Symbiodinium microadriaticum,* for the symbiosis has to be reassembled in each new generation after the host reproduces. Mechanisms for transmission of exosymbioses rely on one partner's ability to locate the other even if they do not live closely together, as in the symbioses between flowering plants and bees, butterflies, and hummingbirds.

But the diversity of solutions to the enigma of heredity could be a strength of symbiotic evolution rather than a weakness. Simplicity in science has a compelling elegance, but as Ernan McMullin, a philosopher and historian of science, admits, simplicity appeals more to our human aesthetic instinct than to any real objective logic.

We need, moreover, to put these symbiotic "flashes of lightning" into perspective. The evolutionary landscape they illuminate is not steeped in darkness but bathed in a pale, shimmering phosphorescent glow, the result of steady, unrelenting Darwinian change. The glow is cut through and altered by these lightning flashes and, quite occasionally, severely shaken and disturbed by the brilliant thunderbolt of a major endosymbiotic event.

The two different evolutionary mechanisms we see in this wonderful landscape are not separate from one another; they are tightly interwoven not only with one another but, in the most astonishing fashion, with the landscape itself.

· 11 ·

PLANETARY EVOLUTION

The tree of life is a twisted, tangled, pulsating entity with roots and branches meeting underground and in mid-air to form eccentric new fruits and hybrids.

— LYNN MARGULIS, *The Symbiotic Planet*

IN 1962 THE SCIENCE HISTORIAN and philosopher Thomas S. Kuhn introduced the concept of "paradigm" to describe how an important new scientific discovery or understanding causes a revolutionary shift in the way scientists think. Examples include Aristotle's *Physica,* Newton's *Principia,* and Einstein's theories of relativity. With the synthesis of Darwin's theory of evolution and Mendelian genetics, neo-Darwinism became such a paradigm. The paradigm concept has important repercussions, for once a paradigm is accepted, it is regarded as beyond challenge.

After the discovery of the chemical structure of DNA in the 1950s, the main thrust of neo-Darwinian argument was closely bound to molecular biology. As genomic mechanisms became clearer — in particular, as they applied to replication, change, and inheritance — evolutionary scientists turned away from nature as a whole and took an increasingly reductionist perspective. In 1976 this view culminated in the publication of Richard Dawkins's *The Selfish Gene,* which advocated that the struggle for existence occurred not at the level of the individual organism but in the most basic unit of inheritance, the gene. This idea, which originated with Dawkins's mentor at Oxford, William Hamilton, coincided with the proliferation of knowledge in genetics. The "selfish gene concept" explained much that had hitherto baffled evolutionists, from the behavior of "jumping genes," which duplicate themselves with a selfish disregard for the genome as a whole, to seemingly altruistic human behavior. No idea could have more neatly encapsulated the modern Darwinian zeitgeist. Dawkins's book achieved such massive popularity that it and Stephen Hawking's *A Brief History of Time* were the only two science books mentioned in a poll of middlebrow British readers asked to list their personal choices of the one hundred most important books of the twentieth century.

The notion of the selfish gene was accepted by the great majority of evolutionary biologists, and it became the central tenet of population genetics and evolutionary ecology. The concept was readily amenable to mathematical extrapolation, and biology teachers treated it as gospel; the next generation of scientists came to maturity assuming that reductionist neo-Darwinism and evolution were one and the same. By the 1970s the mimetic influence of the selfish gene extended into the highly controversial spheres of evolutionary psychology and sociology, where it was applied to individual and social patterns of human behavior.

Nevertheless, an important question remained: was Darwinian gradualism, even with all the panoply of selfish genes, mutation, and selection by nature, sufficient to explain the diversity of life on Earth?

As Kuhn made clear, when a paradigm is accepted not only do scientists stop inventing new theories, "they are often intolerant of those invented by others." It took a rare single-mindedness and courage to dare to contradict the paradigm. However, just three years after *The Selfish Gene* was published, its reductionist focus was challenged by a novel vision, one that looked not to the submicroscopic wonder of the gene but to the Earth in its entirety and, beyond it, to the cosmos.

In September 1965 three people were sitting in a small office in the Space Science Building of NASA's Jet Propulsion Laboratory in Pasadena, California. They were the astronomer Carl Sagan; Dian Hitchcock, a philosopher employed by NASA to assess the logic of their experiments; and the English polymath scientist James Lovelock, who, four and a half years earlier, had been thrilled to work with NASA on plans to explore the moon. In the interim NASA had become more ambitious and had turned some of its attention to Earth's planetary neighbor, Mars. But the Mars program was not going according to plan and Lovelock was convinced that it was based on a very shaky premise. Thirty-six years later, when I asked him about that fateful meeting in Pasadena, he invited me to visit him and his charming wife, Sandy, in their wood-girdled home in southwestern England.

Lovelock had a vivid recollection of that meeting in an office overlooking the San Gabriel Mountains. And he also recalled what he described as a sudden flash of inspiration. A fourth person had joined their gathering, an astronomer named Lou Kaplan, who brought a large sheet of paper with the latest infrared sightings of Mars and Venus that provided a detailed analysis of the planetary atmospheres. Everybody present was looking forward to examining the chemical compositions.

As Lovelock remembered it, "I was even more fascinated because it was already my theory that Mars and Venus were dead planets. Here, suddenly, my predictions were proved bang on. We saw that the atmospheres were full of carbon dioxide and very little else. All the indications were that they were at a chemical equilibrium state. I knew now that there could be no life on either of the planets; if life had been present, the atmospheres would have been changed by it. And so it was at that moment I thought: 'My goodness! How different is our Earth!'"

In time that flash of insight would cause Lovelock to suggest that all of life, from the plankton underlying the marine food webs to the great canopy trees of the rain forest, "could be regarded as constituting a single living entity, capable of manipulating the atmosphere to suit its needs and endowed with faculties and powers far beyond those of its constituent parts."

For a growing number of scientists from many disciplines, this hypothesis, named Gaia, would become the most provocative challenge to the existing perception of our world. Inevitably, from the moment of its birth, Gaia was enmeshed in controversy.

James Lovelock has been variously described as an unorthodox chemist and inventor, space scientist, marine biologist, the Earth Father, and so on. Many such descriptions seem to imply a rebellious nature, but he denies that this is the case. "Not at all. I'm a fairly tribal person who likes to belong. It's just that I was cursed, or blessed as the case might be, with a mind independent enough to see things differently, and so I always found myself in a rebel position."

When he was just four years old, his father gave him a Christmas present of a wooden box full of odd electrical bits and pieces. In retrospect, Lovelock considers this an inspired gift because it encouraged him to think. "Why," the boy demanded, "do you need two wires to send the electricity along when you only need one pipe for water?" Nobody seemed able to give him an answer. "Finally I realized that if I wanted to find out the answers to such questions, I would have to become a scientist and find out for myself."

Eventually he obtained a chemistry degree at Manchester University, then completed his Ph.D. while working at the Medical Research Council in London. Beginning in 1941, he spent twenty years at the council on various investigations, mostly of infection. While studying the common cold at a hospital in Salisbury, he had the opportunity to indulge an innate skill at scientific instrument building. His colleagues, believing that drafts of cold air might play a part in spreading cold viruses, asked him to make an instrument that would measure air currents. Lovelock constructed two anemometers, one incorporating ionization and the other ultrasonics. The ionization version met the need. Ten years later he invented the electron-capture device, a brilliantly simple instrument that can detect trace amounts of pesticides in animal blood and tissues and of chlorofluorocarbons in the atmosphere, which allowed scientists to measure levels of environmental pollution. This device helped vindicate Rachel Carson's timely warning in *Silent Spring,* which alerted the world to the dangers of an environmental disaster.

In typical Lovelockian fashion, he made no attempt to claim the patent on the instrument, which was claimed by the U.S. government in 1964. Money was never the primary motive for this man, who, however eccentric he appeared to his colleagues, was essentially a scientific purist.

It was his inventiveness that first brought him to the attention of NASA. In March 1961 Lovelock was invited to assist them in their plans to explore the solar system. At the Jet Propulsion Laboratory he became a consultant to a team led by Norman Horowitz, a space biologist, whose objective was to devise ways of detecting life on other planets. The team's plan was to dispatch an automated sampling device coupled to a microbiological laboratory to sample the Martian soil and judge if it would support microbes. Other experiments would test for chemicals that normally accompany microbial life.

After a year or so, Lovelock's independent thinking began to show. Disenchanted with the prevailing assumptions, he began to ask himself some questions. Would Martian life, if it existed, reveal itself in tests based entirely on the patterns of life we observe on Earth? Increasingly interested, he probed even more deeply, asking, "What is life, and how should it be recognized?"

When he challenged his colleagues on their current strategies, they asked him what alternatives he was proposing. Lovelock, who had been approaching the problem from the perspective of physical science rather than biology, replied that he would look for a reduction in entropy, based on the second law of thermodynamics, as a cardinal sign of life.

For physical scientists the second law is one of the fundamental planks in understanding the universe. Put simply, it states that when two bodies are brought into contact, heat (energy) flows from the hotter body (higher energy state) to the colder body (low energy state). This may sound like simple common sense, but it has cosmic implications. For example, if the energy of the universe tends to dissipate in the form of heat, then the universe must be winding down from an initial state of enormously concentrated energy: the big bang, as some scientists conceive the start of everything. Life, which gathers and organizes energy — for example, through photosynthesis — appears to reverse this process. But when Lovelock suggested that his colleagues should search for evidence of this, they seemed less than impressed. Nevertheless, the idea took root in his mind. Back home in the quiet of the Wiltshire countryside, he spent some time thinking about the true nature of life.

It wasn't an easy subject to conceptualize. "Like a lot of terms we employ in everyday life, we haven't the slightest idea of what it means."

Lovelock trawled the scientific literature for workable definitions of life, not in conventional biological terms but in physical terms that a chemist could use as the basis for life-detection experiments. His search was unsuccessful. He found a vast repertoire of information on the anatomy, physiology, classification, and evolution of life, but nothing that satisfied his question about its essential nature. With the philosopher Dian Hitchcock, he attempted to probe it further. They asked themselves a much simpler and more specific question: How might a planet be altered by the presence of life?

The only planet they had to go by was Earth, with its variety of living species, its peculiar atmosphere, and its oceans. It was obvious that life on Earth interacts with its environment. For example, some of the raw materials and all of the waste products come from and are returned to the land, atmosphere, or oceans. It seemed likely that the changes they were looking for might show up in these domains.

Applying this new line of reasoning, Lovelock and Hitchcock soon confirmed that Earth's atmosphere, in particular, is dramatically changed by the presence of life. Atmospheric gases are in a permanent state of disequilibrium, which is the opposite of what one would expect from the second law of thermodynamics. A good example is the simultaneous presence of methane and oxygen. In sunlight these two gases should react to form carbon dioxide and water, removing the gases from the atmosphere. Yet the methane level in the atmosphere is relatively stable. To keep it so, vast amounts of methane must be constantly added. The same is true of oxygen, which, though it is being used up in a great many chemical reactions, including combustion and the respiration of animals and plants, stays at a steady level in the atmosphere. Recycled oxygen must be added regularly to the atmosphere to keep its level constant. Lovelock was convinced that the main source of both these gases was life.

Here was simple and yet convincing evidence for his line of thinking that the presence of life alters what one would expect according to the second law of thermodynamics. Moreover, the evidence of this change should be apparent to somebody observing our Earth from a distant source in space. The conclusion was inevitable: "The atmosphere of a life-bearing planet would thus become recognizably different from that of a dead planet."

Lovelock then applied this test to Mars, which, of course, has no surviving oceans but has an atmosphere that can be analyzed from Earth. Even at that time, ten years before the 1976 *Viking* landings, Lovelock was convinced that any life on Mars would have left its signature in the atmosphere. At the meeting in Pasadena in 1965 he learned that the Martian atmosphere consists largely of carbon dioxide. To Lovelock this ruled out the presence of life, a conclusion that was not altogether popular among his colleagues on the NASA project. But Lovelock's mind was already entranced by an extension of his new idea, a grander notion running at a tangent to the Mars deliberations. In his own words, as he reminisced later: "I found, in looking at Earth's wonderful atmosphere, with all of those gases out of equilibrium, yet somehow keeping constant, that something must be regulating this. It was natural for me to think that it was life that was regulating it — and regulating it in such a way as to keep it comfortable for itself."

He went on to examine other gases in the Earth's atmosphere, such as nitrous oxide and ammonia. And when he looked deeper still, through experiments on air pollution, his conclusions differed in spectacular fashion from those of his geochemical colleagues. Lovelock now had evidence that the atmosphere was not some chance mixture of gases resulting from nonbiological processes, as the geochemists assumed. The atmosphere might actually be an extension of the biosphere, intimately interactive with, and regulated by, life.

He needed a name for the concept, and this, famously, was provided by a man who lived in his village: William Golding, a Nobel laureate and the author of many novels, including *Lord of the Flies*. Without hesitation, Golding recommended that this "living Earth" should be called Gaia, after the Greek Earth goddess. It was almost too wonderful a name, for it derived from a second great mind more connected with humanism and the arts and was a conceptual metaphor of huge historical resonance and metaphorical impact.

Now Lovelock needed to convince his colleagues.

In 1969, he was given the opportunity to put forward his Gaia hypothesis at a scientific meeting in Princeton, New Jersey, whose focus was the origins of life. His presentation was not well received. It appealed to nobody except a Swedish chemist, Gunnar Sillen, and a biologist he had never met before who had been given the job of editing the papers presented at the meeting — Lynn Margulis. Lovelock encountered even more resistance when he attempted to have his theory published. Mainstream scientific journals rejected all his submissions. When his paper was sent out for peer review, biologists, in particular evolutionary biologists, treated his ideas with outright scorn.

Even Carl Sagan disagreed with him and, in Lovelock's opinion, not just on scientific merit. "He disagreed with me emotionally. He was so tied up with the American dream of finding life on Mars: it must be there, hiding in some oasis somewhere." When Lovelock addressed a congress of geophysicists at Mainz, Germany, the Europeans among his audience were enthusiastic, one offering to publish his lecture in the journal *Tellus,* while the Americans in the audience were openly hostile.

The more Lovelock promoted his hypothesis, the more this skepticism continued to grow. The geologists in particular were flatly dismissive. As the eminent geochemist H. D. (Dick) Holland of Harvard University exclaimed, "We don't need Gaia. We can explain all the facts about the Earth through straightforward geology and geochemistry." The microbiologist W. Ford Doolittle of Halifax, Nova Scotia, made the comment: "If Gaia were to be real, it would require trade unions of all the species of organisms to meet annually on Mount Ararat and negotiate next year's climate." Richard Dawkins was more focused in his criticism: life forms could not regulate anything other than their own kind. Evolution, viewed from a reductionist neo-Darwinian perspective, could not lead to cooperation on a global scale. Dawkins considered Gaia's webs of mutual dependency mere pop ecology. Natural selection meant that nature had to be presented with choices to select from.

Lovelock later conceded that the neo-Darwinian criticisms were the greatest challenges for his theory. Yet he believed that Gaia and natural selection were perfectly compatible.

In the early 1990s John Maynard Smith went so far as to refer to Gaia as an "evil religion." Such attitudes, as much as the refusal of prestigious scientific journals to publish any papers related to Gaia, wounded Lovelock. "I was puzzled," he confessed, "by the response of some of my scientific colleagues who took me to task for presenting science this way . . . It is the scientific establishment that now forbids heresy." Frustrated by years of academic resistance, he published his hypothesis as a book for a general audience, *Gaia: A New Look at Life on Earth,* in 1979. In it he wrote: "I see the Earth as more than just a mixture of living things and inanimate matter. I see it as a tightly coupled entity, where the evolution of the living things and the evolution of the inorganic matter constitute a single and totally inseparable process. It's a whole system."

It is ironic that even as his planetary vision was under ferocious attack, the Gaia theory, unbeknownst to Lovelock, was in actuality a step further along an older road of scientific exploration.

Belief in the balance of nature goes back to the Greeks and has been incorporated into many of the great systems of religious belief. But all such holistic thinking had been attacked by Darwinian-minded evolutionists and ecologists since the early decades of the twentieth century. Despite the opposition, belief in the essential interdependence of life had never entirely gone away. Unfortunately, the concept did not become any easier to define, given the labyrinthine complexity of life, and any "natural balance" was difficult to conceptualize and even more difficult to subject to rigorous scientific testing. An apparently insoluble drawback, in the words of Frank N. Egerton of the University of Wisconsin at Parkside, was that "a synthetic balance-of-nature theory would necessarily have included knowledge from earth and atmospheric sciences as well as from botany and zoology."

In the nineteenth century, many scientists adhered to the older scientific philosophy that knowledge could be advanced only by the systematic accumulation of facts. Theoretical deduction — first constructing a hypothesis or theory that could subsequently be challenged — was not their way of doing things. But Darwin, for example, took a different approach, examining the world through a preformed conceptual framework. In a letter to a friend, dated September 18, 1861, he reflected on the advances in geology since his voyage on the *Beagle.* "About thirty years ago there was much talk that geologists ought only to observe and not theorize; and I well remember some one saying that at this rate a man might as well go into a gravel-pit and count the pebbles and describe the colors. How odd it is that anyone should not see that all observation must be for or against some view if it is to be of any service."

In 1926 a Russian mineralogist and biogeologist, Vladimir I. Vernadsky, asked himself whether the landscape evolves in an intimate partnership with life. Eschewing any prior hypothesis, Vernadsky set out to advance knowledge through a series of objective observations that, once pointed out, become self-evident to any thoughtful observer.

At this time the evolution of life was seen as a serendipitous phenomenon that had no relationship to the inanimate world. Unconvinced by such Darwinian reductionism, Vernadsky believed that a critical evolutionary relationship existed between life and its geophysical environment, a relationship he later encapsulated in the concept of "biosphere," a name that had earlier been proposed by Professor Eduard Suess, of Vienna University, for the geological zone in which life dwells. Vernadsky's vision was dramatically grander, redefining "Biosphere" (spelled with a capital B) to encompass the "envelope of life where the planet meets the cosmic milieu." For Vernadsky, Biosphere included both life and its environment: the Earth's surface, oceans, and atmosphere. In its interaction with the landscape, life was not merely a geological force, it was *the* geological force that had shaped the surface of our planet over the eons of evolution. And since it derived its energy from the sun, Biosphere, and the diversity of life within it, was a cosmic phenomenon that could be understood by the same laws that applied to such constants as gravity and the speed of light.

Of prime importance to Vernadsky's thinking was the exchange of gases involved in respiration, which on the one hand was vital for the evolution of all forms of life and on the other determined the composition of the atmosphere. Perhaps his greatest contribution to our understanding of life and its interaction with the inanimate Earth was his recognition of the central importance of the steady and essentially eternal supply of solar energy, which life converted and reutilized through photosynthesis. This primal transfer of energy from cosmic source to planetary recipient makes possible many of the great cycles of life.

For example, the oxygen in the atmosphere is produced by photosynthesis, notably by microbes and planktonic life forms in the oceans. But as Vernadsky perceived, these photosynthesizers do not actually "create" this oxygen; they liberate it from oxides "as stable and as universal as water and carbon dioxide." Thus does life interact with inanimate nature, in the process radically altering the chemical composition of the Earth's crust and further changing it through the highly reactive nature of the released oxygen. Today we are well aware that carbon dioxide, produced as a byproduct of the respiration of oxygen, has an important greenhouse effect on Earth's surface temperature. Vernadsky popularized the need to take measurements, however colossal, to quantify these vital processes.

He estimated that the total mass of "green" life responsible for the trapping of sunlight and the production of atmospheric oxygen was 738×10^{15} grams of organic carbon, which is nearly a hundred times the mass of the life forms dependent on that oxygen. The global mass of rock-dwelling lichens amounts to a staggering 13×10^{13} tons, a biomass greater than that of all life in the oceans. These colorful symbiotic life forms, which are photosynthetic through their algal partners, wear the rock face away to soil, which in its turn is enriched by the humus of dead organic life, to become the root habitat of plants. Similarly for every environment, and every life form, Vernadsky dissected out and measured chains of interactions. In the oceans alone, he estimated the organic carbon content (mainly photosynthetic life forms) to be about 1.9×10^{12} tons.

Vernadsky observed many other interactions that form part of what we now regard as the cycles of life. He examined the biological processes involved in the formation of limestone and the numerous interactions that make up the cycles of food and waste disposal, including the organic cycling that is an essential part of death and decay. From the planetary perspective, all such cycles consist of a transfer of solid and liquid matter between life and the environment. Water itself erodes rock and alters the landscape. It dissolves minerals washing in from the land and ferries these to the seas. And water has its own natural cycle, evaporating from the oceans, carried in clouds, then returning to the land as rain. Even in the form of rain, water cycles minerals, another facet of the great interactions of Biosphere.

In this way, Vernadsky measured and confirmed not only the great cycles of interactions, in which life forms depend on or cooperate with other life forms, but also a colossal series of interactions between all of life and the inorganic environment. Life, he explained, generates biomass, biological reproduction, and physical movement, radically changing the composition of the Earth's crust, whether soil or rock or water, and, most vital, the atmosphere. Life directly controls all crustal geological processes. In a clear echo of earlier beliefs in the balance of nature, Vernadsky concluded that the biosphere as a whole should be viewed as a single orderly mechanism, its formation influenced by life at every step, and subject to fixed laws.

This vision was revelatory, but for more than half a century after publication of his book in Russian, and its translation into French three years later, Vernadsky was ignored by the Western world. As Lovelock admitted, "When I first formulated the Gaia hypothesis, I was entirely ignorant of the related ideas of scientists such as

Vernadsky." Lovelock made a great intuitive leap beyond Vernadsky's observationally based deductions in deciding that the biosphere evolved its own systems of self-regulation. But if he was to convince the world, he desperately needed the help of an evolutionary biologist courageous enough to do battle with paradigms. This he gained with the growing support of the redoubtable Lynn Margulis. Gaia and symbiosis, two quite different and distinct paradigms, acquired a curious symbiosis of their own.

Lovelock and Margulis developed one of those interactive relationships that happen now and then in science, in which innovative minds from disparate disciplines come together. In Lovelock's words, "Lynn's deep knowledge and insight . . . was to go far in adding substance to the wraith of Gaia." She began by helping him redefine Gaia as "the series of interacting ecosystems that compose a single huge ecosystem at the Earth's surface, incorporating all of life, the atmosphere, oceans and soil." The totality could be viewed as a physiological system, including temperature, alkalinity, acidity, and reactive gases, equipped with feedbacks that keep the physical and chemical environment constant and in a state comfortable for life.

Years earlier, in ending his employment with the British Medical Research Council, Lovelock had abandoned any notion of permanent employment, whether in academia or industry. "It gave me a feeling there were tramlines going all the way to retirement and the grave . . . I thought, 'To hell with it! I'm going to break free.'" It was a tough decision for a married man with a wife and four children to support, but he was able to fund his own researches through a much extended scientific career in which he made a living from his inventive dexterity and an extraordinary capacity for innovative thought.

Confounding his unrelenting neo-Darwinian critics, Lovelock declared that his original definition of Gaia, as put forward in his first book, was wrong. He had come to accept the Darwinian dictum that organisms cannot regulate anything but themselves. Where earlier he had claimed that life adjusted its environment to suit itself, the newer

Gaia theory proposed an intrinsic coupling of life with its inanimate environment that in the totality was self-regulating.

Lovelock now put forward a model analogy in the form of a world similar to Earth in which the only plants were black daisies and white daisies. The model is more complex than I explain it here, but essentially, if the temperature is too cold, dark daisies dominate, because they absorb more heat and thrive, and if the temperature is too hot, the dark daisies suffer from overheating while the white daisies thrive, by reflecting the excess heat back into space. Using this "Daisyworld" model, Lovelock demonstrated, with elegant simplicity, how the planetary temperature could be regulated by the competitive growth of dark- or light-colored plants. Lovelock did not pretend that Earth was this simple: the purpose of Daisyworld was to answer the criticisms of Ford Doolittle and Richard Dawkins, who claimed that Gaia was not objective but teleological. In a more complex argument based on Daisyworld modeling, Lovelock confirmed that it was not life alone but the whole system of life tightly coupled with the physics and chemistry of the Earth's environment that does the regulating, the process of regulation being an emergent property of this tightly coupled system. To satisfy the neo-Darwinians, he even introduced a capacity for cheating, with gray daisies given a slight advantage over the others. However, the gray daisies did not proliferate and conquer. Computer extrapolation of his Daisyworld model confirmed that it allowed autoregulation of climatic temperature. But if Lovelock thought these steps would satisfy his critics, he was mistaken. As recently as 1998 Gaia was dismissed in contemptuous terms by the Darwinians D. Robertson and J. Robinson in the *Journal of Theoretical Biology*.

In a sense such antagonism was predictable, for Gaia theory has been the greatest challenge to the prevailing reductionist thinking. In his autobiography, *Homage to Gaia,* Lovelock concedes, "If Daisyworld is valid, seventy-five years of neo-Darwinian science will need to be rewritten."

To the late Stephen Jay Gould, professor of geology and zoology at Harvard University and a leading American evolutionist, Gaia was "warm and fuzzy and it strikes a chord . . . a pretty metaphor, and not much more." John Maynard Smith disagreed with Margulis's support for Lovelock. "I believe she's wrong about Gaia . . . But I must say, she was crashingly right once, and many of us thought she was wrong then, too." Even Ernst Mayr declared, "It's startling to find a reputable scientist arguing such fantasies."

But were these scientists right in their opposition? Surely there is something very curious about the long-term stability of the Earth's environment. Geologists and astronomers have questioned why Earth has kept its oceans while Mars and Venus are dry. When ecologists, basing their calculations on Darwinian assumptions, predict that with increasing complexity, ecosystems become more fragile, why, we might ask, are such complex ecosystems as the rain forests wonderfully stable? Such observations suggest a resilient and sustained homeostatic stability.

In spite of the rancorous opposition, Gaia theory is now moving from the shadows of unorthodoxy into the light of contemporary consideration. For the science journalist Fred Pearce, writing in *New Scientist,* Lovelock is to science what Gandhi was to politics. "His central notion, that the planet behaves as a living organism, is as radical, profound and far-reaching in its impact as any of Gandhi's ideas." Indeed, in the opinion of the Swiss historian of science Jacques Grinevald, Gaia "is the major cultural and scientific revolution of our time."

There will continue to be debate and differences of opinion between believers in Gaia theory and Darwinian-oriented evolutionists, but few would disagree with Lovelock's emphasis on the importance of a holistic view. "It is little use having climatologists, earth scientists, ocean scientists, community ecologists and atmospheric chemists all working in separate subjects, separate buildings, and worse, with no common language between them." We have seen how the lack of common ground has dogged evolutionary science, leading to misunderstandings between Darwinians, symbiologists, and the proponents of Gaia theory. At the same time, if we are to examine the contributions of Darwinism, symbiosis, and Gaia theory to the wonder of life's evolution on Earth — indeed if we are to consider how these forces interact with each other — we need to dispel any residual confusion about symbiosis and its evolutionary role. Since the concept of symbiosis was first mooted by de Bary, enormous advances have been made in our understanding of biology and genetics. It would seem both appropriate and reasonable to reexamine symbiosis from this modern perspective.

· 12 ·

REDEFINING CONCEPTS

> If you analyze the evolution of a viral lineage, you discover that 80 percent of the genes have no counterpart in the genetic database. What this means is that viruses are capable of creating complex genes all by themselves. The oceans are filled with viruses like these, so what you have is genetic creativity on a very large scale, a kind of biological big bang.
>
> — Luis Villarreal

When Charles Darwin put forward his theory of evolution, he compared deliberate selection by humans — in the breeding of animals or crossing of grains for better yield — to what must also happen in "nature." To conceptualize this, he used the metaphor "natural selection." Darwin made no attempt to consider evolutionary change arising from interactions between species: he limited his theory to nature selecting the "fittest" individuals within a species. His theory also demanded a mechanism of inherited change, since nature required some degree of choice for selection to operate.

As we have seen, neo-Darwinism assumes that the mechanisms of genetic change — gene mutation and recombination of genes as part of sexual reproduction — are random and that the creative force is natural selection. A literary analogy makes the implications clear. Let us say that a writer merely jots down random thoughts as they come into his head, leaving it to the editor to make sense of the script. The editing, which is the essential role of natural selection for Darwinians, is more creative than the act of writing.

Symbiogenesis, the evolutionary force that derives from the interaction of different species, reverses this situation: the creative force is not natural selection but the act of symbiotic union.

The genome is, of course, the sum total of all the genetic material that makes up the heredity of any given life form. We have seen that this heredity is not confined to the nucleus but is shared between the nucleus and certain organelles within the cytoplasm. The study of genes is called genetics, and the study of how genomes work within themselves and how they compare with one another is called genomics. Both genetics and genomics have advanced at astonishing speed, opening up a new world that is of profound importance to the understanding of evolution. It is also a world in which symbiogenesis can be seen as a major evolutionary force, deriving from the essential nature of the genome, with its pluripotent capacity for change.

I propose to show that the genome has an additional capability above and beyond those of self-replication and template function. Implicit in its chemical structure — or "chemical behavior" — is the innate capacity for change.

In every example of evolution, genomic change always precedes natural selection. Smith and Szathmáry are in no doubt about this: "The crucial step [in evolution] is to understand the mechanism of heredity, because the whole process of evolution by natural selection depends on it." In his metaphor of the selfish gene, Dawkins also directs attention to the genetic level. The genome is the ultimate determinant of life. The physical bodies we recognize — in plants, animals, fungi, protoctists, even bacteria — are all constructed around the template of their genomes. What natural selection does is select the consequences of genomic change that has already taken place when, for one reason or another, the physical expression of that change gives the organism some advantage for survival and, implicitly, better reproductive fitness. Any alternative theory of evolution, such as symbiosis, must, by implication, also govern and be governed by the genome.

In de Bary's original definition of symbiosis as "the living together of differently named species," he embraced parasitism and predation but excluded short-term associations. This definition has proved important and enduring, but its vagueness has given rise to what David Lewis, emeritus professor of botany at Sheffield University and a leading British authority on symbiosis, has described as "soggy semantics." Some interpret de Bary's insistence on long-term associations as excluding such mutualisms as the pollination of flowering plants by insects and hummingbirds. Such exclusions make no conceptual sense, since these involve interactions between different species with obvious evolutionary significance. Lewis has suggested that the concept of symbiosis merely requires that two organisms interact with each other. In his words, "It requires neither close nor prolonged association."

Lynn Margulis subdivides the evolutionary force of symbiosis (symbiogenesis) into four categories of increasing intimacy: "ecological," in which the interaction involves behavioral sharing; "metabolic," in which a chemical product or metabolite is shared; "gene-product transfer," in which a protein or RNA molecule is donated to a partner; and complete "gene transfer" between partners. I will adopt both Lewis's and Margulis's views and add in a third: the genomic perspective.

Werner Reisser, emeritus professor at the Botanical Institute at the University of Leipzig and a lifelong symbiologist, has long been interested in symbiotic green algae. In 1992, with the role of viruses in mind, he suggested a change to de Bary's definition, from the "living together as different species" to the "interaction of dissimilar genomes." With some slight redefinition to include the thinking of Lewis and Margulis, I propose the following genomically based definition:

> Symbiogenesis is evolutionary change arising from the interaction of different species. It takes two major forms: endosymbiosis, in which the interaction is at the level of the genomes, and exosymbiosis, in which the interaction may be behavioral or involve the sharing of metabolites, including gene-coded products.

This definition will provide the basis for my discussion of the evolutionary role of symbiosis throughout the remainder of this book. Because symbiogenesis is rather a mouthful, I will revert to the simpler term "symbiosis" as an understood abbreviation for it. By genome I mean all of the hereditary material, that is, the DNA and RNA in the cytoplasm as well as the nucleus. The symbionts do not need to live together over the long term since this makes no difference from an evolutionary perspective. Moreover, from this genomic viewpoint, the definitions of endosymbiosis and exosymbiosis are somewhat different from their usual counterparts. The powerfully creative force of endosymbiosis is now given full emphasis: it is a mechanism for sudden evolutionary change arising from the fusion of genomes, whether this involves the transfer of a single gene or the blending of the genomes of different life forms. In this sense endosymbiosis equates with the fourth and most intimate of Margulis's categories.

The shared behavior of the hermit crab and the anemone is a typical example of exosymbiosis, which also includes the gut-living microbiota of ruminants and the root-living fungi of mycorrhizae. While endosymbiosis may be less common than exosymbiosis, it is much commoner than Darwinians assume.

In this definition of symbiogenesis I have avoided the term "fitness," which would be inappropriate since, as Kwang Jeon discovered, endosymbiosis does not necessarily improve survival or reproductive capacity. It could, indeed, be deleterious to the survival of one or more of the interacting partners, or to the resulting hybrid organism, the "holobiont." What symbiogenesis does is to create genetic change. Some fine-tuning of the symbiotic partnership will still be necessary, but if we attribute this role to natural selection we are employing a very different, indeed rather emasculated, concept compared to Darwin's. In symbiosis, the decisive evolutionary event is the interaction between life forms and not the subsequent honing of the organism.

My definition will help resolve some dilemmas. For example, the pollination associations involving insects, hummingbirds, and flowering plants are now readily accommodated as exosymbioses. It also clarifies the distinction between endosymbiosis and exosymbiosis in cases that might otherwise cause confusion. For example, a bacterium that lives within the tissues of its host is no longer endosymbiotic, as some would currently classify it, but exosymbiotic, unless we can demonstrate a direct interaction between the two genomes. It will also help resolve a dilemma that I consider to be the most important of all: the symbiotic evolutionary potential of viruses.

Biologists differ on whether or not viruses constitute living entities. It all comes down to one's definition of life, and, as Lovelock discovered, this is not as simple as it seems. Margulis, for example, takes the view that they are not, because they lack the chemical metabolism necessary for their own living processes. Having personally worked with viruses, both in the laboratory and in treating infected patients, and having sounded out the opinions of many other experts, I take a different view. For me, viruses are every bit as alive as a butterfly or a great blue whale.

Like all other life forms, viruses are coded by genes, made up of DNA or RNA. They are classified by strains (the equivalent of species), grouped within lineages called families. They replicate using genomic mechanisms similar to those of all other life forms. Viruses even have their own complex behaviors, for example, in penetrating the host's immune defenses, in locating the target organ and cell within the host while evading the fierce attack of antibodies and cellular immunity and, in the case of pandemic influenza viruses, in spreading throughout the world without any means of locomotion. I have taken this argument further in *Virus X,* but here it might be useful to summarize my views in the form of a thought experiment.

Let us suppose that our purpose is to create an organism that will survive and prosper in the strangest, most wonderful ecology on Earth: the living genome. Consider the characteristics this organism will need to possess. Physically, to inhabit what is essentially a molecular world, our hypothetical organism must be ultramicroscopic — as indeed viruses are, being on average about one one-thousandth the size of a bacterium. To accommodate the physiological and biochemical requirements of life on such a minuscule scale, the organism can take advantage of its environment, which is nothing other than a factory of genomic programming and manufacturing capability. In Darwinian terms, the organism adapts to its ecology.

Viruses are wonderfully adapted for survival within the genomes of their hosts. Indeed, if we do not regard them as alive, how do we deal with the awkward fact that all the laws of evolution apply to entities that are not living? Every viral infection involves a meeting of genomes. When the host is a single-celled life form, such as a bacterium or protist, or a germ cell in an animal, plant, or fungus, all biologists, whatever their views on viruses, agree that this interaction has evolutionary potential.

We are now ready to examine the mechanisms that underlie the broad sweep of evolution. In doing so we shall consider where and when the great evolutionary paradigms interact not only with each other but also with the inanimate world, as proposed by Lovelock's Gaia.

And where better to start than the very beginning, the first glimmer of life on Earth.

PART II

The Weave of Life

The study of evolution is vast enough to include the cosmos and its stars as well as life, including human life, and our bodies and our technologies. Evolution is simply all of history.

— LYNN MARGULIS, *The Symbiotic Planet*

· 13 ·

THAT FIRST SPARK

Shall we conjecture that one and the same kind of living filament is and has been the cause of all organic life?

— ERASMUS DARWIN, *Zoonomia* (1794)

ON SUNDAY, February 12, 2001, two rival organizations, one privately sponsored and one public, announced simultaneously that they had completed the first comprehensive analysis of the human genome. The work leading to this accomplishment had been going on for years, and the breakthrough, though monumental, did not mean a complete understanding of how our genes and chromosomes make us human, any more than it really explained the differences between individuals. It was nevertheless remarkable for what it did reveal.

A major surprise was the relatively modest size of the human genome, about 40,000 genes. For many scientists as well as nonscientists, this small number was humiliating. It had been assumed that our species, with all of its complexity, would have more than 100,000 genes. To put it into perspective, we have only ten times as many genes as a bacterium, four times as many as a fruit fly, and twice as many as a nematode worm. It seems that we are not, at least in quantity of genomic memory, vastly more complex than these humble life forms, though it could be argued that our genome is more complex in the way in which its genes interact with each other. Subtlety may be more important than numbers, or, as reporters Philip Cohen and Andy Coghlan argued in *New Scientist,* it's not how many genes you've got but what you do with them that counts. For many people, perhaps, the most revealing aspect was the confirmation of how much we have in common with every other form of life on Earth. For example, we share 2,758 of our genes with the fruit fly and 2,031 with the nematode worm; and the three of us — human, fly, and worm — have 1,523 genes in common.

I doubt that Darwin would have felt humbled by this news. I rather think he would have been exhilarated, because he would have realized that this shared inheritance could not have arisen by chance. It is incontrovertible proof of his intuitive belief that all of life on Earth has a common ancestry. And this confirmation, in turn, lifts the lid on a box full of mysteries. Where did such life come from? When did it arise? How, from such humble beginnings, did life come to populate every ecological nook and cranny of the Earth? Are the forces that created life so universal that its evolution is inevitable throughout the universe? Or is there something peculiar to Earth and its solar system such that we really are alone?

These ground-zero questions have obsessed the human imagination for as long as people have enjoyed the gift of reason.

They certainly inspired the imagination of Francis Crick, the codiscoverer of the stereochemical structure of DNA. In 1981 he published an influential book, *Life Itself,* in which he put forward his view that life could only have come from outer space, seeded on Earth by some extraterrestrial intelligence. Crick is not the only great scientist to propose an extraterrestrial source of life. The renowned British astronomer Sir Fred Hoyle and the mathematician Chandra Wickramasinghe have long proposed that life must have originated outside our solar system. Hoyle believed that life forms from space are constantly arriving on Earth and mingling with our biosphere. These extraterrestrial theories have evoked a rare unanimity among symbiologists and Darwinians, who condemn them as explaining the origin of the specious rather than the species. But are such theories as far removed from reality as their detractors claim?

If current theories are correct, our solar system formed from the collapse of a cold, dark interstellar cloud, which evolved into a vast, swirling disk of fiery gas and dust before condensing out as our sun and planets 4.6 billion years ago. The dark cloud itself must have been preceded by the colossal explosion of a supernova: heavy elements such as the iron that colors our blood red could never have formed in the nuclear fusion of modest suns such as our own. In the most evocative way imaginable, the myths of our pagan ancestors have been confirmed: life on Earth, including we humans, was born in the furnaces of the stars. Our moon may have broken off from the still-condensing Earth after a titanic collision with another planet the size of Mars. From the same primordial effluence came the iron magma that became the molten core of our planet.

In this first phase of its evolution, known as the Hadean period (after Hades, from the Greek for "hell"), Earth, still cooling from the huge energies of its origins, was an extremely violent place. Throughout the first half billennium of its existence, comets and asteroids subjected it to frequent bombardment. Even now, in a continuing legacy from those cosmic beginnings, the Earth's gravity draws up hundreds of tons of dust and meteorites every day, and this cosmic dust is rich in organic molecules formed in that same dark cloud that spawned our solar system. Recently, particles collected under the wings of specially adapted aircraft at altitudes of 62,000 feet, have confirmed that this material contains as much as 50 percent organic carbon, a higher percentage than is found in any other known extraterrestrial object. Meteorites, even more than comets, have been found to bear an astonishingly rich variety of organic compounds, including those that form the chemical basis of DNA.

The water, gases, and complex organic chemicals necessary for the beginnings of life may well have come from space; on the other hand, everything needed to spark off the evolution of life may have been present on Earth, derived from its beginnings in that dark interstellar cloud. Certainly this was the view of Charles Darwin, who on February 1, 1871, wrote a letter of personal reflection on this subject to his contemporary Joseph D. Hooker:

> If (and oh, what a big if) we could conceive in some warm little pond with all sorts of ammonia and phosphoric salts — light, heat, electricity present, that a protein compound was chemically formed, ready to undergo still more complex changes, at the present day such matter would be instantly devoured, or absorbed, which would not have been the case before living creatures were formed.

Though he might have envisaged an evolutionary stage involving organic chemicals, Darwin's ideas were nothing more than inspired guesses. He had no knowledge of DNA and its remarkable properties of self-propagation. Moreover, the first spark of life owed nothing to either Darwinian or symbiotic evolution. This germinal step or, more likely, a vast number of trial-and-error steps, involved much more elemental forces.

In the atomic structure of molecules are forces of attraction and repulsion for other elements and molecules that would have been crucial to the beginnings of life. Paul Davies, a physicist and science writer based in Adelaide, Australia, argues that even the known forces of classical physics and chemistry cannot explain this prebiotic stage of evolution. He looks to a more complex explanation involving information theory coupled with quantum mechanics. Whatever the ultimate explanation, it seems likely that elemental forces of a physicochemical nature made possible this first stage of evolution. Amino acids are the building blocks of proteins. Although we know that these are manufactured readily from natural processes, a vital step in this prebiotic stage had to involve a template molecule that coded for how various amino acids would be strung together to form proteins. The template molecule also needed to find ways of preserving its own integrity while evolving a means of self-replication that permitted biological reproduction. With the exception of certain viruses, which are based on RNA, all of life fulfills these two roles through DNA. An important stage, therefore, had to be the evolution of RNA and DNA molecules.

We may never know what those ancestral chemicals were, but two main fields of thought exist. Some argue that the evolution of a self-replicating DNA-like molecule must have taken place independently of the evolution of protein manufacture, and the two then came together in a single protocell. Others favor a single evolution, almost certainly involving a preliminary phase based on RNA. The reliable replication of a DNA molecule of any substantial length requires the presence of enzymes. But at the same time, the enzyme has to be coded for by a DNA molecule of a certain length and complexity. In this we encounter a chicken-and-egg situation, since the enzyme itself would have required a lengthy period of prior evolution. To confront this dilemma, we need to understand a little of the basic structure of DNA and RNA.

In the same way that proteins consist of strings of amino acids, rather like letters adding up to a very long word, DNA and RNA are strings of nucleic acids. But where proteins have an alphabet of twenty letters, DNA and RNA have only four. Three of these are the same for both DNA and RNA — guanine (G), adenine (A) and cytosine (C) — but in DNA the fourth is thymine (T), and in RNA it is uracil (U). DNA stands for deoxyribonucleic acid, which gets its name from its component sugar, deoxyribose; RNA, ribonucleic acid, is named for ribose. This is all one needs to know to grasp the fundamental coding of life. Every gene is nothing more than a long word made up from a sequence of the four letters. A specific triplet of letters codes each of the twenty amino acids, so the coding of DNA translates directly to the coding of proteins.

The seemingly small difference — the substitution of thymine for uracil — between DNA and RNA has considerable physicochemical significance. Although DNA codes for proteins, its three-dimensional structure is radically different from that of a protein, while the structure of RNA more closely resembles a protein. Enzymes, of course, are proteins. In 1967, the scientists Carl Woese, Francis Crick, and Leslie Orgel proposed that RNA molecules might act as their own enzymes. In the early 1980s Thomas R. Cech and Sidney Altmann confirmed that this was the case, winning the Nobel Prize in chemistry for their discovery of the first RNA-based enzymes. This discovery is of great evolutionary importance, for it means that RNA has the potential to catalyze its own replication. The Harvard biologist and Nobel laureate Walter Gilbert then proposed that before the evolution of DNA as the central molecule of life and heredity, there must have been an "RNA world."

It is intriguing to note that the genomes of many viruses are based on RNA, and some evolutionary virologists believe that those viruses may have evolved during this RNA phase of evolution.

In the early 1950s it was assumed, from radio telescope observations of primordial matter, that the atmosphere of the Earth 2 to 3 billion years ago contained large quantities of hydrogen, water, ammonia, and methane, a mixture that came to be known as the "primordial soup." In 1953 the American chemist Harold Urey attempted to test some protolife hypotheses by reproducing this primordial soup in his laboratory. Teaming up with a twenty-two-year-old graduate student, Stanley Miller, at the University of Chicago, Urey filled a glass flask with these chemicals, then passed a series of electrical sparks through them to simulate the energy-donating properties of natural lightning. Over the following week, Miller and Urey found that the water in a connected flask turned brown. When analyzed, the flask's contents were found to include amino acids.

This experiment was hailed as providing the first evidence for a prebiotic phase of evolution.

Some of Miller and Urey's assumptions were wrong. We now know, from analysis of certain sedimentary rocks, that the atmosphere of the ancient Earth was much simpler, containing water, nitrogen, carbon monoxide, and carbon dioxide, and that from about 2 billion years ago, it also contained increasing amounts of oxygen. But the mistaken assumptions were not damning. The Miller and Urey experiment did confirm that natural processes could readily convert simple organic chemicals into those important for life.

We also know that deep under the oceans, around the fissures where tectonic plates meet and volcanic lava erupts, an environment is found in which sulfides precipitate out as filmy chemical membranes and even form gelatinous bubbles that might have provided a "capsule" for evolving protobiotic chemical precursors. By and large, the most ancient bacteria still found on Earth grow best at temperatures close to the boiling point of water. All of these data point toward life originating in hot sulfurous springs or deep in the seas, around the vents along the midoceanic ridges, where volcanic activity brings boiling water saturated with sulfides to the surface.

If amino acids were easily manufactured from basic ingredients, the natural origin of DNA and RNA are more problematic. Various scientists have presented scenarios to explain how these nucleic acids might have arisen and afterward were strung together to form the first self-replicating strings. From this stage onward evolutionary forces came into play.

Let us say that there are two competing strings: string A, requiring an abundance of the chemical *a*, and string C, requiring an abundance of the chemical *c*. Assuming that a supply of raw materials is critical for replication, if *a* is more abundant than *c*, string A will manufacture more of itself — in Darwinian terms, "nature" will "select" string A for survival. The first priority, from the evolutionary perspective, is individual survival; the next priority is reproduction. Whether or not the daughter strings of A share its "fitness" will depend on accurate and abundant reproduction. If there were no constraints, the self-replicating strings would double in number with every generation, eventually overwhelming the ecosystem. In reality, limited supplies of the necessary chemicals prevent this result. If there are many variant strings in that ecosystem, each variant must also compete with the others, so that over very long periods of time, fewer strings will succeed.

Chemical evolution of this sort has been tested experimentally in simple noncellular replicating systems, and this pattern of selection has inevitably ensued. Even at the very threshold of life, a prototype form of natural selection emerges as a key force.

As those early strings copy themselves, errors crop up. Mutation, the underlying hereditary mechanism of neo-Darwinism, is also present at this earliest stage. And this is mutation on a grand scale. Even the master molecule DNA — which is much further along the evolutionary scale than these early progenitors — is prone to errors. Without control and repair enzymes, our own genes would make one error in every two hundred replications of the DNA letters. The mutation rate of the early replicators would have been much greater even than that of HIV viruses, which are known to give rise to a great diversity of mutations in a single AIDS sufferer if untreated.

At a very early stage along the evolutionary pathway, certain of those replicating strings discovered a new wonder.

Consider that by chance two very different strings, A and B, come together. Just as it is a basic property of DNA-like molecules to reproduce themselves, it is also a property of these strings to join. The union results in the string AB. A new proto–life form is born, not through mutation but through the interaction of two protogenomes. This is the earliest equivalent of endosymbiosis: two entire strings, each with its individual evolutionary history, have come together to form a holobiont, which must take its chances of surviving in competition with the parent strings.

Mathematical models express the survival statistics of this mutual give-and-take. For instance, if A behaved like an aggressively selfish virus, the survival potential of the holobiont population would be decidedly poor. But now and then the composite string, AB, has a better survival potential than A or B alone. Kathleen H. Keeler, of the department of ecology and evolutionary biology at the University of Nebraska, has deduced some elegant mathematical extrapolations of this type of mutualistic association, with cost-benefit models for survival.

The amalgamation of several strings into one, which then replicates as a unit, can take place only if natural selection selects for cooperation among its constituent units rather than selfish competition. In the words of Smith and Szathmáry, "The origin of the [DNA] code assumed cooperative interactions between replicators . . . It is clear that, for life to progress beyond the stage of simple replicating molecules to something more complex, a web of cooperative interactions between genes had to evolve." Even at this primal level of evolution, the creative change arises independently and *de novo* out of a genomic potential — in this case endosymbiosis — but long-term survival, and further mutational honing of the holobiont string, requires the intervention of natural selection.

A genomic perspective even at this primal stage clarifies the relationship between the two great evolutionary forces.

One essential improvement along the tortuous evolutionary route to life must have been the acquisition of stability during replication. Without stability, advantageous characteristics would soon be lost to replication error and the protolife string would become extinct. Another improvement — believed by some scientists to have been the greatest symbiotic union of all — may have been the coming together of the template-forming mechanisms of DNA and RNA and the protein-forming mechanisms based on amino acids. This was the hypothesis put forward by Freeman Dyson, emeritus professor of physics at the Institute for Advanced Study in Princeton, as the "double origin" of life.

Nobody pretends to know the intricacies of these earliest stages. We must assume that the interplay between the proteins that the self-replicator coded for and the evolving genome became increasingly complex, resulting in some form of self-organizing chemistry within an enclosing membrane, to produce the first protocellular structures.

One implication of this strange and wonderful story of early evolution is that, because the raw materials necessary for that first spark are ubiquitous, the potential for life is written into the elemental forces of the universe.

This interpretation is very different from that expressed by the Nobel Prize–winning French biologist Jacques Monod in 1962, when, reflecting the views of the majority of scientists of his day, he concluded that the evolution of life on Earth was such a chance event that we must be alone in the universe. Today our knowledge of the forces involved in life's beginnings has grown, and another Nobel Prize–winning French biologist, Christian de Duve, has expressed the opposite opinion: that given the right conditions, the evolution of life "is a cosmic imperative." Many scientists at NASA now share de Duve's "biological determinism," and they are actively searching for alien life in our solar system and, indeed, throughout the universe.

If these NASA scientists are right, we have already found the first evidence of such alien life.

· 14 ·

THE TRUE PIONEERS

"This organism is as good a candidate as you could want for something arriving on a Martian meteorite." It survives on hydrogen, carbon dioxide, and nitrogen given off by seabed volcanic vents. Given these conditions, "it wouldn't be unexpected to find something like this elsewhere in the universe."

— ROGER HIGHFIELD, quoting John Reeve

DAVID GILICHINSKY works at the Institute for Biological Problems in Soil Science, near Moscow. After many years of drilling into the permafrost in northern Siberia looking for mammoths and other long-preserved animals, he has shifted his interest to bacteria. Working with NASA, his team has found evidence to suggest that bacteria endured prolonged freezing at temperatures of −20° Celsius for perhaps 8 million to 15 million years. In this extraordinary discovery we glimpse an essential requisite for survival in Earth's early history.

After the Hadean period, during which the spark of life struggled for existence in a barren world of evolving chemicals, Earth moved into a new period known as the Archaean (from the Greek word for "ancient"), during which the first true bacteria appeared. To survive in the harsh conditions then prevailing, these bacteria developed endurance unequaled by any other form of life. Today scientists have found certain of the ancient bacteria known as archaebacteria flourishing in the sulfurous hot springs of Yellowstone National Park and others inhabiting the stygian darkness of rocks miles underneath the

Earth's surface, where they "breathe" iron rather than oxygen. The ancient bacterium *Haloarcula marismortui* is one of two species that have discovered the secret of living in the Dead Sea, and *Thiobacillus thiooxidans* can digest concrete, a property that may soon help engineers clean up sites of radioactive contamination.

These examples suggest that similarly enterprising life forms could have evolved elsewhere in the universe, and on August 7, 1996, NASA scientists told the world in dramatic fashion that they had found evidence for this in a Martian meteorite.

Mars, as the fourth planet from the sun and, with Venus, closest in orbit to the Earth, is a prime candidate for the evolution of extraterrestrial life. There are other reasons for anticipating life on Mars, at least at some stage in the past. Its gravity is less than, though not extremely different from, that of the Earth, and it has similar durations of day and night, although the year is almost twice as long, 687 days. However, as Lovelock and his colleagues discovered, the Martian atmosphere differs greatly from that of Earth, with 95 percent carbon dioxide and only one one-thousandth as much water. Most experts agree that the thin atmosphere, the dry, oxidizing soil, and the constant surface exposure to ultraviolet radiation from the sun would make it extremely difficult for life to begin on Mars as it is today.

But could life have started there when conditions were more conducive billions of years ago? The visible surface of Mars suggests that much earlier in its evolution the planet was warmer and that its surface may have been washed by gigantic oceans. Its two moons, Deimos and Phobos, might have dragged tides over those long-lost oceans. In this climate of speculation, David McKay, head of the NASA scientific team, claimed to have found evidence of fossilized bacterial shapes in a Martian meteorite.

The meteorite had been collected in 1984 in Antarctica. Shaped like a small, pockmarked potato, it was believed to have been chipped from Mars 15 million years ago, when an asteroid or a comet crashed into the planet, hurling pieces of rock into space. The debris was caught up by Earth's gravity about 13,000 years ago. McKay revealed to an attentive audience from all over the world that microscopic analysis of the meteorite revealed tiny spheres and rods that had the typical appearance of bacterial fossils. "All the facts point to the simplest explanation — that these are evidence of early Martian life."

The announcement provoked an inevitable backlash. Skeptics pointed out that NASA had every reason to engage public interest in Mars, for it needed to persuade the U.S. government to fund the highly expensive missions to that planet. But McKay's team remained adamant about the claims. Using sophisticated microscopes, together with laser-based chemical analysis of the meteorite's microstructure, they demonstrated three corroborative features. The sizes and shapes of the tiny fossil forms were similar to those of known bacteria on Earth. And in close association with the bacterial shapes they found iron-containing crystals of a sort produced by certain bacteria. Finally, they determined the presence of carbon-based organic molecules never found before in any Martian meteorite.

But the skeptics were not convinced. It was argued that the bacterial forms were too small to be true bacteria, that the carbon content might be from contamination on Earth, that the shapes were nothing more than a natural clustering of rocky crystals. NASA could not claim that the results were conclusive, although it was prepared to offer pieces of the meteorite to other scientists for analysis.

Attention focused on the minute examination of magnetite crystals found within the meteorite. These were shown to be extremely pure and defect-free, which in general would point to a biological origin, though it was possible that the pollution-free environment of Mars, unaided by bacterial life, could have created them. Proponents pointed to the length of the crystals' string structures as indicative of a biological origin, since magnetite crystals of nonbiological origin tended to collapse into blobs of indistinct material. Other supporters thought that the forms represented not bacteria as such but the chemical "footprints" of bacteria, tiny vesicles that certain bacteria produce within rock crystals.

The debate revived a far more controversial hypothesis. If bacteria did once evolve on Mars, could those bacteria have been transported to Earth, not as fossils billions of years old but as living cells, which then played an important part in our own evolution? This ancient idea, known as panspermia, was given a logical basis by the Swedish chemist and Nobel laureate Svante A. Arrhenius in 1908, when he proposed that radiation from stars could blow living bacteria from one planet to another. The Martian meteorite reawakened interest in such theories. Restating the central conviction of their book, Hoyle and Wickramasinghe found it inconceivable that life on Mars and Earth would have followed such a similar route simply by chance. "Our hypothesis," claimed Wickramasinghe, "is that life is a cosmic phenomenon."

Scientists had long been aware that bacteria could survive environmental extremes by forming "spores," in which the genome effectively hibernated within a tough enclosing shell, ready to spring to life again when conditions improved. Nobody quite realized the extremely high protection this might confer until Russell Vreeland, a microbiologist at West Chester University in Pennsylvania, isolated a hitherto unknown species of bacterium from salt crystals 560 meters down in the salt beds of Carlsbad Cavern in New Mexico. Still viable within tiny globules of water sealed within the crystals were bacterial spores dating back an estimated 250 million years, before the time dinosaurs first roamed the earth. Taking great care to avoid contaminating his test material with modern bacteria, Vreeland liberated the spores from hibernation; to everybody's astonishment, they grew into living bacteria on growth media.

Of course, Vreeland's results will have to be corroborated by similar findings elsewhere in order to be accepted. We need to be especially cautious in considering theories that life on Earth comes from other planets. If NASA's findings prove correct, however, all evolutionary theories will have to take account of the existence of life at a more cosmic level.

On one point no controversy exists: that remarkable endurance allowed bacterial life to begin astonishingly early in Earth's history. Microfossils found in ancient rocks from Australia and South Africa confirm the existence of bacteria 3.5 billion years ago. These fossils tend to be found in rocks laid down from the mud around hot springs in a volcanic landscape. And even older rocks from Greenland, dated at 3.8 billion years, contain the fingerprints of carbon-based chemicals that could only have belonged to some primitive bacteria.

An earlier generation of biologists would be astonished to learn that life could have evolved barely 100 million years after the geological cooling of the Earth, when the violence of its formation was barely subsiding. Whatever rudimentary forms of life first appeared in such a setting, their existence must have been precarious.

The planet was still being bombarded by debris from space. Torrents of volcanic rage dictated the weather. The atmosphere would have been similar to that of Venus and Mars today, rich in carbon dioxide and nitrogen and devoid of oxygen. A single great landmass was surrounded by a vast ocean in which tidal waves were triggered by the still cooling moon. The landscape would have appeared utterly barren to our eyes, with no plants and no bird or insect life. The water would not be recognizable as the green-blue seas we know today. It was full of natural pollution, including abundant amounts of elemental iron that could only remain in solution in the absence of oxygen. This is the cradle in which life began.

Until recently, biologists recognized only bacteria they could study in laboratory cultures. Over more than a century, this approach, pioneered by Louis Pasteur and Robert Koch, had defined a mere 5,000 bacterial species. As a consequence, there was a profound ignorance of the evolution of bacteria. Then, in the 1970s, Carl R. Woese, a microbiologist at the University of Illinois at Urbana, together with his colleague Norman Pace, of Indiana University in Bloomington, pioneered new genetic techniques in their search for unknown species of bacteria. Scattered through the cytoplasm of cells are structures known as ribosomes, which manufacture proteins. A kind of RNA, ribosomal RNA, conducts the actual assembly line. Pace and Woese focused on a gene, known as *16s* rRNA, that is part of the ribosomal RNA. This gene is present in every known organism on Earth and has played a fundamental role in the evolution of life.

The importance of the *16s* gene lies in the fact that it mutates so slowly that its nucleotide coding sequence remains identical over vast periods of time. The *16s* ribosomal RNA in a human is the same as that of a dog, in spite of the many millions of years since the evolutionary divergence of humans and canines. Put another way, when you do find differences in this ribosomal gene, the life forms you are comparing are far apart in their evolutionary journey — certainly as far as, and often much farther than, a mere difference of genus.

In the 1970s, Pace first used *16s* to look for new bacteria, and almost immediately he discovered large numbers of previously unknown varieties. Soon other biologists adopted this genomic approach and identified a great diversity of new life forms. As John Holt, a microbiologist at Michigan State University, commented: "You could go out in your back yard, and if you really put your mind to it, you could find a thousand new species in not much time." Vigdis Torsvik, a Norwegian biologist, did just that with an alternative genetic probe. She calculated that a single gram of soil contained something on the order of 10,000 species of bacteria, twice as many as had been found in the preceding hundred years.

Other scientists have extended this approach to a wide variety of ecologies, always with the same results. In samples of open ocean, Jed Fuhrman and his colleagues at the University of Southern California found sixty-one microorganisms, only four of which were previously known. In Obsidian Pool, a hot spring in Yellowstone Park, Pace and his graduate student Sue Barnes discovered sixty new species. So extensive and diverse are the new discoveries that scientists are now wondering if there might be hundreds of thousands, possibly millions, of bacterial species, the great majority still unknown to science.

Woese found such dramatic genetic differences between types of bacteria that he proposed a radical shakeup of the taxonomic divisions of life. While Lynn Margulis disagrees with certain aspects of Woese's classification, the consensus of opinion is that bacteria represent life's true pioneers. They conquered the inhospitable wilderness of the primeval Earth with no other mechanisms than the genomic capacity for change coupled with nature's selectivity.

Since no life form had gone before them, these autotrophic bacteria were obliged to construct every metabolic pathway from the raw materials of an inanimate environment. Taking in water, nitrogen, carbon, oxygen (though not as a gas, since the atmosphere was anoxic), hydrogen, and sulfur — as well as smaller quantities of many other elements, such as iron and phosphorus — these bacteria solved the great survival enigmas that faced them, manufacturing a huge complexity of novel organic compounds that would subsequently prove essential to all of life. Exposure to direct ultraviolet light from the sun could be lethal, as could a slight change of acidity or alkalinity, a sudden swing of climate, the arrival of toxic rain, or desiccation.

In Death Valley, California, the temperature of exposed ground can go as high as 190°F, and the surface is bone dry and caked with toxic salts. But when Ken Nealson, a NASA-based scientist interested in the possibility of life in arid worlds, cut into the ground, he discovered a triple sandwich of colored layers. Just beneath the surface was the bright green of photosynthetic algae; next, a red layer of bacteria; and at the bottom a brown-black layer of sulfur-metabolizing bacteria. The green layer captured the energy of sunlight, while the other two accepted the first layer's metabolic wastes as raw materials and reciprocated with the products of their own metabolism — a primal example of symbiotic specialization.

Other autotrophic bacteria colonized worlds deep beneath the oceans, under arid dust on land, or within grains of rock and ice. Every advance in colonization by bacteria, from the macroecologies of equatorial heat and polar cold to the microecologies of rock pools and soil strata, marked a new evolutionary step. One adaptation was the ability to ferment sugars and other carbohydrates into compounds containing less energy, such as alcohols. This allowed bacteria to extract carbon from the carbohydrate molecules while storing the liberated energy for other essential chemical reactions. Symbiotic partners took the alcohol waste and extracted still more carbon-rich energy from the molecules, thus playing a part in what would become one of the great cycles of Gaian geophysiology, the carbon cycle. Other bacteria learned how to use elemental nitrogen and sulfur, and one by one the great cycles of life were born.

All subsequent life, including our human species, owes an immense debt of gratitude to these bacteria, which evolved a great many of the metabolic pathways still central to our living chemistry.

Scientists were determined to find out how this spectacular achievement happened. In the mid-nineteenth century, biologists recognized that bacteria have no gender, so that all offspring derive from a single parent. During the reproduction known as binary fission, the bacterium swells up, copies its circular molecule of DNA, and then divides in two, creating daughter cells that are clones of the parent. The whole process usually takes about twenty to thirty minutes. Another means of achieving the same end is for the bacterium to form, from its enclosing membrane, a small "daughter" bud that swells, takes up a copy of its replicated DNA, and grows to full size before separating from the parent.

But changes can take place in bacterial genes, for example through mutations during replication of the DNA, thus altering the daughter's genome. These mutations arise in roughly one of every million divisions, a frequent occurrence given the fact that a single bacterium dividing every twenty minutes would, with unlimited space and raw materials, produce 2^{288} copies of itself in just four days — more than the number of protons or of quarks that physicists have estimated to exist in the universe. Of course, space and resources severely limit this almost infinite potential, yet the number of divisions is still enormously high, making mutation a very significant genomic force for change. Although most of these mutations are useless or even harmful, occasionally one results in some small improvement in the function of the coded protein. Constantly molded by natural selection, the effects can accumulate over time.

We see an example of the power of mutations today in the typical winter outbreaks of influenza. The accumulation of mutations in the viral genome makes each new strain of flu sufficiently different that our immune system fails to recognize it as the same virus that came around last year, and so a great many people suffer from flu annually. There remains, however, a considerable commonality with all the previous flu viruses, so our immune system's fightback is partially primed, and the illness is not usually too serious. The situation is quite different in pandemic flu, which results from two different strains of influenza virus swapping whole chunks of their genomes during infection of a single host (usually a pig). This wholesale genome-to-genome interaction, essentially an endosymbiotic event, creates an entirely new strain of the virus. That is why the Spanish flu pandemic of 1919 carried the terrible lethality of a true emerging plague, for the human immune system no longer recognized the virus.

Mutation is also a significant feature in the current pandemic of TB. In every tuberculosis sufferer, among the billions of bacteria present at the beginning of the illness, there are many mutants, some of which are bound to be resistant to each known antibiotic. This is why tuberculosis so readily becomes resistant to the drugs used in its treatment and why it can be cured only by giving three appropriate drugs at once.

Bacteria have many other mechanisms of change. On October 19, 1946, Joshua Lederberg, then twenty-three, and his mentor at Yale, Edward L. Tatum, published a very brief paper — only a quarter page long — in the British journal *Nature*. Their report changed our understanding of bacterial evolution, bringing together the disciplines of genetics and microbiology. Up to that moment, biologists had assumed that bacteria reproduced only by binary fission, in which genetic variation was limited to what could arise from mutation. Lederberg and Tatum showed that bacteria from two different strains (the equivalent of species) could exchange genes. Although scientists interpreted the finding as showing that bacteria have sex, in fact it gave them a potential that was far greater than that. The implications for microbial genetics were so important that Lederberg and Tatum were subsequently awarded a Nobel Prize. The potential for change that they identified, now called genetic recombination, is common to all sexually reproducing species and is the basis of the modern science of genetic engineering. For microbes, particularly bacteria, recombination allows for very rapid evolution.

Recombination can take place in several ways. One of these is a kind of sexual process in which bacteria of two strains come into contact and, through cytoplasmic bridges, exchange DNA. In another, bacteria take up DNA fragments from other strains even when they are dead and their DNA has been released into the environment by the dissolving of their bodies. In yet another form, DNA enters the bacterial genome through an infecting virus known as a bacteriophage. If you share my view that viruses are life forms, this is an endosymbiotic union of genomes. In modern genetic language, these mechanisms are often collectively referred to as "lateral" or "horizontal" gene transfer. All examples of lateral gene transfer are, by definition, sudden interactions between genomes. When they involve genomic interaction between different species, they are by definition endosymbiotic.

Yersinia pestis is the bacterium that causes bubonic and pneumonic plague, which was responsible for the terrifying series of epidemics known as the Black Death that repeatedly decimated huge areas of the Eurasian landmass up to the seventeenth century. The causative bacilli have not gone away; there have been sporadic outbreaks of plague in Asia, Africa, and the American Southwest, including an epidemic in India in 1994. In 2001 a group of scientists from the Pasteur Institute in Paris, working with colleagues in Antananarivo, Madagascar, reported the transfer of resistance to the drug streptomycin from one plague bacillus to another, a capacity *Yersinia* shares with bacteria of several different species. This transfer of resistance is mediated by a mobile genetic unit, or plasmid.

The evolutionary potential of lateral gene transfer is so enormous that Sorin Sonea, a professor of microbiology at the University of Montreal, has suggested that all the strains of bacteria in the world form a single pool in which "temporary, adaptable symbioses" that solve problems of survival are the way of life. This exchanging of genes has led to a common "information bank," which makes the designation of individual strains or species a good deal less meaningful than is the case with eukaryotic life forms. The individual strains of bacteria that make up this gestalt can be thought of as the cells of a "global superorganism" that has evolved over billions of years of endosymbiotic fluidity.

This proposal might appear confusing, but it clicks readily into place if we consider that these different forms of genetic recombination and the progressive accumulation of mutations are all manifestations of a single creative force: the pluripotent genome. In this way, the genetic mechanisms underlying Darwinism and symbiosis are readily accommodated under a single conceptual umbrella. For simplicity's sake, I shall refer to this all-embracing concept as the "Genome," with a capital G. Natural selection still plays a part in symbiotic evolution, but it is only that of fine-tuning the symbiotic partnership after the creative act of symbiotic union. And here, in the interaction between the evolving organism and the ecology, Gaia also enters the picture.

As Lovelock has shown, all life forms interact with and change the ecosystem. For instance, the metabolic activities of bacteria change the rock, soil, water, or air in which they live. If the changes enhance the organism's fitness, it prospers. If the changes pollute its ecology, thereby reducing fitness, the organism declines. A small, local change that improves fitness for one life form can, by degrees, influence other life forms in the same ecosystem through the interplay of parasitism, predation, food webs, and the myriad possibilities for exosymbiosis and endosymbiosis. In Sonea's view, the bacterial superorganism alters the environment in which its component strains evolve and colonize. This, as he suggests, "is a perfect example of the Gaia hypothesis of how life modifies the environment."

The importance of the age of bacteria was first revealed in 1930 when a geologist named G. H. Chadwick proposed a twofold division of Earth's history. He termed the part of the geological record in which animal fossils are visible in rocks the Phanerozoic, from the Greek *phaneros,* meaning visible, and *zoíon,* meaning life. This period extended from 570 million years ago to the present. The immense expanse of time before visible fossils he termed the Cryptozoic eon, or "age of secret life," which reigned for 2, or even 3, billion years.

The Cryptozoic was a time of immense change, during which minuscule life forms — bacteria, almost certainly in association with their viruses — colonized every virgin habitat and terrain, directly incorporating the elements of earth, air, and water into their pioneering of life.

Some of the archaebacteria took hydrogen for their metabolic needs directly from the atmosphere. Others took it from the hydrogen sulfide gas belched from volcanoes and other geothermal vents. Then some new bacteria came on the scene, the first green life forms, formerly known as blue-green algae but now more correctly designated cyanobacteria. Their arrival was of formative importance, for now the evolving Genome had found the means of trapping the energy of sunlight. Discoveries made in Greenland, on the edges of the polar ice cap, have revealed the chemical signature of photosynthesis some 3.9 billion years ago, suggesting an astonishingly early date for this milestone in evolutionary history. As Margulis and Sagan described it: "In an evolutionary innovation unprecedented, as far as we know, in the universe, the blue-green alchemists, using light as energy, had extracted the hydrogen from one of the planet's richest resources, water itself . . . This single metabolic change in tiny bacteria had major implications for the future history of all of life on Earth."

The processes of life can be seen as a complex weave of dynamic processes, driven by chemical energy. Through photosynthesis, life discovered a consistent and sustained source of energy, leading to the stores of energy we call fats and carbohydrates. Energy storage made it possible for other life forms to evolve that did not need to photosynthesize for themselves.

The oxygen released by photosynthesis became a waste product that diffused back into the surrounding oceans and ultimately into the atmosphere. For the great majority of bacteria, a new danger loomed: oxygen was poisonous to them. Faced with extinction, they had two options. Either they had to find new habitats away from the poisonous oxygen — deep in mud or in the intestines of animals or miles under the surface of the ground — or, better still, they had to evolve a new means of surviving.

The creative Genome answered this challenge: other bacteria evolved the ability to breathe oxygen. One of these was the forerunner of our mitochondria. Like the bacterial colonies that eke out a living in Death Valley, the cyanobacteria formed sandwich layers of symbiotic collectives in shallow coastal waters, bonding with inorganic particles of sand and debris to form mushroom-shaped rocks known as stromatolites. These strange shapes can be seen in the shallow water of Shark's Bay, off the north coast of Australia. Little by little, they spread until the entire shallow ocean of the Proterozoic Earth was studded with colonial stromatolites.

Then the Genome, powered by the dual forces of photosynthesis and oxygen-based respiration, added still more spectacular innovations to its growing library of change. Bacteria began to fuse genomes to form more complex protoctists. One of the oxygen-breathing bacteria entered into a new endosymbiotic union with the evolving protist and became its mitochondria.

Change, massive and irrevocable, was literally in the air.

· 15 ·

THE GREENING OF THE EARTH

> Then God said, "Let the earth bring forth vegetation: every kind of plant that bears seed and every kind of fruit tree on earth that bears fruit with its seed in it." And so it happened.
>
> — GENESIS 1:11

WHAT A STRANGE and barren place the Earth was in Precambrian times, before the valleys were carpeted with meadows of grasses and wildflowers, before the arrival of the first rudimentary shrubs or trees. It was not, of course, devoid of life. Bacteria must have betrayed their hardy presence as intriguing patches of color — though not the colors we associate with life today. Then, with the arrival of the cyanobacteria, a great change began to wash over the land. In certain places, along the coastlines of brackish lakes and seas, spreading maps of green appeared. A primal spring had arrived on Earth, a spectacular triumph for this almost miraculous ultramicroscopic entity, the evolving Genome, and selected — taken up with wide-flung arms — by nature herself.

In Lovelock's metaphor, Gaia had taken her first breath.

Certainly that breath was a slow, hesitant gasp, as we would expect of a newborn, but all of the wonderful forms and colors we associate with nature would follow from it.

A myriad of living interactions followed, at first erratic and local, but gradually reaching out into the atmosphere and oceans, shaping not only the continuing evolution of life itself but the face of nature, from the mountaintops to the oceans, from the equator to the poles. The endosymbiotic fusion of bacteria gave substance to the first eukaryotic cell, its all-important replication apparatus cocooned within the double membrane of its nucleus. The blue color of Earth's sky, the "speckled sapphire" of a living planet that took away the breath of the *Apollo 8* astronauts who first saw the dawn of their planet from the moon, results from the volumes of oxygen breathed out by photosynthesis.

At every evolutionary step, large or small, the creative Genome continued to experiment. Through intrinsic genomic mechanisms coupled with selection, mitochondria and chloroplasts were honed and edited, many of their genes transferred from the endosymbionts to the nucleus, leaving behind the husklike organelles we see today in the cytoplasm. Far and wide the Genome experimented, using every creative force in its arsenal for change, throwing out an ever-increasing variety of new templates, each competing with its parents and rivals for survival and, ultimately, leading to further change.

As Darwinians such as Richard Dawkins, Stephen Jay Gould, and John Maynard Smith have attested, mutation does not necessarily lead to increasing complexity. But every endosymbiotic change enlarges the Genome, inevitably increasing the complexity of its interactions. A proliferation of bacterial species gathered around the photosynthesizers, discovering in their waste a new source of nutrition, and this stimulated further Genomic experiments. Some changes took place at the submicroscopic level and involved the most endosymbiotic of all organisms, the viruses, while on a larger scale, though still microscopic, an amoeboid protist was infected by or ingested a cyanobacterium. The infection or ingestion failed; instead, a new endosymbiosis resulted in the incorporation of the cyanobacterium as a chloroplast, leading to the forerunner of all plant cells: the first green alga.

The similarities among mitochondrial genomes seen today suggest that all of them come from a single ancestor. But in chloroplasts, the range and diversity of genomes are wide. Many photosynthetic life forms have formed novel symbiotic unions with eukaryotes, suggesting that in the trials and errors that proliferated during early evolution, symbiosis was far from simply the living together of friendly life forms. Indeed, symbiosis is not about Mr. Nice Guy, who comes along and shakes hands with Ms. Nice Girl, and everything is hunky-dory from then on. Symbiosis is about predation and parasitism and, even in its mutualistic form, it is about tough bargaining and hard compromising, with each partner's survival depending on the outcome. Many such bargains and compromises did not survive, simply disappearing without a trace from the record of life. It is, however, a testament to the evolutionary advantages of symbiosis that many others did survive, to give rise to the wide diversity of lineages seen today.

Endosymbiotic union with photosynthesizers has given rise to three orders of green algae as well as a wide range of single-celled organisms, notably the flagellates and the tiny creatures known as phytoplankton, which play a vital role in the food webs of the oceans. Within the kingdom of the fungi, photosynthesizers have entered into a great variety of symbiotic unions, such as the lichens. At this time the Gaian system of feedback and balance began, with implications for land, oceans, atmosphere, and climate. One result was the invasion of dry land by the kingdom of plants, which had already begun to evolve in the oceans. Most authorities now believe that the first land plants appeared about 420 million years ago.

As late as the Cambrian period, 570 million years ago, the land, either bare rock or naked mud, was still hostile to life. It is no surprise, therefore, that the first plant life evolved in water. Fossils that look like marine algae have been found in rocks dating back 3 billion years. In February 2000, the botanists Claude Lemieux, Christian Otis, and Monique Turmel, at the University of Laval, Quebec, threw light on the next great evolutionary step when they unraveled the chloroplast genome of a single-celled alga called *Mesostigma viride* and showed that it was the closest known living relative to the common ancestor of all green plants.

The surface of *Mesostigma* is covered with square scales; asymmetrical in configuration, its forward end is equipped with a single "eye spot." Like a speedboat, it propels itself through water in the hunt for food, its twin flagellae powered by the engine of its chloroplast partner. These elementally delightful creatures are among those that Margulis has removed from the kingdoms of plants and animals and assigned to the kingdom Protoctista.

The Canadian discovery offers us a glimpse of life around the fringes of lakes and oceans 800 million years ago, when the ancestral algal strain split into two lineages. The first lineage was the streptophytes, which gave rise to the land plants (the streptophytes' closest living algal relatives, the charophytes, are still abundant today in calcareous springs, for example in Derbyshire); the second was the chlorophytes, which gave rise to the marine forms and all other green algae. By Cambrian times the chlorophytes had already evolved to plantlike forms similar to many of the seaweeds today, with individual cells already specialized to carry out distinct functions.

Off the coast of Australia, swimmers with snorkels weave through willowy kelp forests, some varieties of which can grow 30 meters tall. The weight of the kelps, if dragged onto dry land, would be considerable, yet in their natural habitat they require no support other than the buoyancy of the water. Seaweeds vary greatly in shape, size, and color, from the brown bladder rack and laminaria that wash up on seashores around the world to the bright green lettuce seaweeds (*Ulva* species) and the brilliant red and delicately arborescent *Corallina officinalis,* which waves its fronds over deep coral reefs. Seaweeds contain specialized tissues, so we tend to think of them as complex plants, but most of the varieties we come across on the shore are colonial algae that do not reproduce using sperm and egg to form embryos. Like *Mesostigma,* they are not classed as plants but as members of the phyla Phaeophyta, Rhodophyta, and Chlorophyta within the kingdom of the protoctists.

The colonization of dry land by plants has been described as "the most dramatic event in their history." Although they had developed photosynthesis to supply energy, they also needed to evolve mechanisms of extracting minerals from soil, a much more difficult process than the ready diffusion of minerals into their cells in water. As the paleontologist and biologist partnership of Mark and Dianna McMenamin demonstrate in their visionary book *Hypersea,* perhaps the greatest need of all was water itself. On dry land, subjected to the whims of climate and season, desiccation was an ever-present danger. The first steps in overcoming these obstacles came from a series of symbiotic unions with fungi.

Fungi evolved at much the same time as algae in the primal aquatic environment. But the two life forms are very different in their processes of nutrition. Whereas algae synthesize their food using the energy of sunlight, fungi degrade organic materials on or below the surface of the soil to absorb minerals, in the process obtaining a little water, which is augmented by the rainfall dispersed around grains of soil and rock. The access to groundwater would prove so crucial for the evolution of land plants that it is not surprising that 97 percent of modern land plants use symbiotic chloroplasts for photosynthesis and symbiotic fungi for water and mineral absorption from the soil.

One of the earliest symbiotic unions of algae and fungi created the lichens, which include some of the hardiest life forms on Earth. Recent discoveries have put the origins of lichens back to 400 million years ago, and possibly even earlier. Today they are the main plants of the tundra, where they provide food for reindeer, and in many high-altitude habitats they are almost ubiquitous. They grow on rocks, the bark of trees, undisturbed soil, and even concrete, tolerating extremes of temperature and desiccation. The enduring success of this symbiotic life form can be judged from the fact that they have diversified to 13,500 species, incorporating approximately forty different genera of photosynthesizing algae. Biologists have found lichens eking out a precarious living in the most inhospitable environment on Earth, Antarctica's rocky valleys. In these frigid landscapes, swept by polar winds and devoid of animals, plants, and even insects, they endure temperatures of −324°F by burrowing into the crystals of excoriated rocks. They cluster close enough to the surface for a meager trickle of sunlight to penetrate; their endosymbiotic algae manage to photosynthesize when the temperature rises to 1°F. This illustrates an important role for the life-enhancing strategies of symbiosis; when the going gets tough, there is strength in cooperation.

Another obstacle during the early evolution of plants was the need to maximize their capture of sunlight. The buoyancy of water had allowed seaweeds to assume a variety of sizes and shapes and to distribute their leaflike fronds without any need for special support structures. Minerals and gases in the water could diffuse through the surface of the frond. But on land, plants had to find ways to stand against gravity, to lift and spread their canopies wide to the energy-giving sunlight. This demanded the evolution of roots, stems, branches, and twigs — and ultimately leaves. The larger and more complex land plants became, the more they needed new mechanisms to distribute nourishment and water, to exchange gases with the atmosphere and to reproduce and disperse their offspring. True sexual reproduction, in which the genomes of two parents fuse to create a single offspring, had evolved 600 million years before the movement onto land. But now the sperm, or male gametes, of these terrestrial newcomers could no longer take advantage of the buoyancy and natural movement of water to float to the female gametes, or ova.

David Read, in the department of botany at the University of Sheffield, England, is a world authority on mycorrhizae. As he illustrated for me during an interview about the beginning phase of land plant evolution, the early types, such as *Cooksonia,* were very similar to some contemporary ferns in the genus *Psilotum,* which are little more than rigid stalks devoid of leaves or true root systems. The only surface adornments were slightly projectile scales with bulbous tips, which contained the reproductive spores. Fossils of *Cooksonia* have been found in New York State, where extensive tidal flats periodically flooded and dried out as the water levels of the shallow Silurian seas fluctuated. Read explains, "Those first plants began in water. They didn't need roots and they didn't need to resist desiccation. They already had cellulose. As soon as they arrived on land, their exposed stems needed to resist the drying effects of open air." This resistance they achieved by the development of a hard outer layer, the epidermal cuticle, which covered all of the exposed shoot. An important evolutionary development, the cuticle prevented the loss of water, but it also stopped gases from entering and leaving the plant. To allow the exchange of gases, such as oxygen and carbon dioxide between the plant cells and the surrounding atmosphere, small openings, called stomata, had to evolve.

Another key step was an efficient management system for water. The McMenamins have constructed an elegant metaphor for the symbiotic partnerships that enabled water to flow through all of terrestrial life. At the beginning of their book they acknowledge their debt to Vernadsky, who, in his conceptualization of Biosphere, recognized not only that life was the great geological force but that, because water was so important, life could be viewed as "animated water." That image gives a certain frisson to the ancient derivation of both the Scotch "whisky" and the Irish "whiskey" from the same old Celtic-Irish root, *uisce beatha* (pronounced "ishkeh baha"), which literally means "the water of life."

Plants pump life-giving water out of the ground and distribute it to the most distant twig, flower, fruit, and leaf, much the way our mammalian hearts and circulatory systems pump and distribute blood through oxygen-exchanging lungs to every organ, tissue and living cell of our bodies. The system in plants (and animals) is called a vascular system, and the plants that employ it are called vascular plants. In essence, the message of *Hypersea* is that the blood of these plants is precious water.

Most larger plants today are vascular: the roots, assisted by symbiotic fungi, absorb water and minerals, which are carried up into the plant through special tubes that distribute them to every living cell. A second vascular system carries the nutritional products of photosynthesis down from the leaves for distribution to all of the cells. Water loss is kept to a minimum by a number of mechanisms, the most important of which is the tough waterproof armor of skin, the epidermal cuticle, which seal-wraps all vulnerable exposed surfaces.

If we were to visit the Silurian landscape of 420 million years ago, we might catch a glimpse of the first true land plants around pools and swamps. By the end of the Silurian, vascular tissues had developed, although still in a primitive and mechanically weak form, with specialized tissues near the heart of the stem and support tissues surrounding them. These early plants were low and creeping, lacking properly developed stems and roots. A very early example is *Rhynia,* which had a single protoroot that just flopped down on the muddy ground. From this primitive root grew leafless stalks resembling chives, about 17 centimeters high, that were brilliant green from the abundance of photosynthetic chloroplasts. Like ferns and other spore-bearing plants we are familiar with today, they needed a damp habitat for at least part of the year, so the sperm could travel to the egg in moist conditions.

Little by little, as fungi-plant symbioses evolved and diversified, and as mutations modified forms and metabolism, the complexity of these early plants increased.

The first mycorrhizae would have enabled the symbiotic fungi to forage around the roots for phosphorus, which occurs in very low concentrations, two or three orders of magnitude lower than other elements necessary for plant growth. Infection by fungi may also have induced a defensive reaction in the plant, leading to lignification, the woody evolution that gave the stems and, later, trunks and branches, a supportive stiffening that helped them stand against gravity as well as develop vascular tissues. Six different types of mycorrhizae evolved over the next 100 million years in adaptation to varying climates and soils, though only four types are commonly found today. In the majority of these the flow of nutrients goes both ways, with water, minerals, and phosphates, in particular, moving from fungus to plant, and with carbon compounds flowing from plant to fungus.

Today one form of mycorrhiza, the "vesicular-arbuscular" variety, is abundant in tropical forests, savannas, and temperate grasslands. At least 75 percent of all vascular plant species benefit from this particular union, including many ferns, conifers, tropical rain forest trees, and most flowering shrubs and smaller plants. The main site of nutrient exchange is the arbuscule, a densely branching fungal intrusion found in the root, giving it a hairlike appearance. Recognizable arbuscules have been found in the roots of *Rhynia,* which formed low marshes in the Silurian period. These looked like mangrove roots lying flat on the ground with tubular stalks branching skyward from the single fat root.

Another resident of ancient landscapes left its fossil mark in what is known as the Rhynie chert, a black flint formation in Aberdeenshire, Scotland. For paleobotanists, this has proved to be one of the best fossil-bearing strata in the world as a result of the seepage of water from nearby silica-rich springs. The Rhynie chert fossils include amazing examples of cooperative living 400 million years ago. For example, chytrids, which are varieties of protists, are found inhabiting the tiny bodies of fossil algae, which themselves inhabit the stems of plants.

Many other fossils allow a glimpse into this primal world, in which the first large amphibians were foraging around the shores of nearby rivers and lakes, and the first dragonflies were fluttering in the branches of great lycopod trees, with their asparagus-like leaves spiraling around the trunks. The trees of these first swampy forests were assisted in their struggle to survive by the arbuscular mycorrhizae. The extent of these forests may be judged by their fossilized graveyards, which we exploit today in the form of coal.

Although there are some plants, such as the brassicas, sedges, and rushes, that do not depend on mycorrhizae, 97 percent of all plants today are mycorrhizal, and 80 percent of these are so dependent on their fungal symbionts that they could not survive without them. It is hardly surprising that mycorrhizae have been called "the universal symbiosis." Of course the cells of both the fungus and the plant involved in every mycorrhizal partnership contain symbiotic mitochondria, while the green branches and leaves contain symbiotic chloroplasts.

A recent discovery has shown how Darwinian-driven selection pressure may have given rise to the first broad leaves. During the Devonian period, which began 408 million years ago, many primitive plants acquired spiny leaves that grew directly out of their stems, thereby increasing the photosynthetic surface. Recently, David Beerling and his colleagues at the University of Sheffield realized that the evolution of the much more effective broad leaves we see today coincided with a 90 percent drop in atmospheric carbon dioxide 360 million years ago. The photosynthetic stems of early plants had few stomata, which facilitate gas exchange with the surrounding atmosphere and allow the transpiration and evaporation of water, thus helping to keep the plants cool. The evolution of broad leaves allowed a great increase in photosynthesis, but plants with few stomata would have overheated because of poor transpiration and evaporation. Beerling's team found that broad leaves became possible only because of a hundredfold increase in the numbers of stomata, an adaptive response to the falling levels of carbon dioxide during this period. The broader leaves, coupled with more efficient root systems for providing water, still managed to stay cool and moist.

As the complexity and diversity of mycorrhizae spread, another common variety became abundant in temperate and boreal forests, especially those dominated by beech and pine. Roots bearing these fungi are easily recognized: they have few or no "hairs" and tend to be short and thick. The root apex is enclosed in a fungal sheath from which the hyphae extend into the soil. On high ground and in more northern latitudes, another form of mycorrhiza partners members of the heath family, including the heathers, azaleas, rhododendrons, bilberries, blueberries, and huckleberries. A fourth variety of mycorrhiza made possible the evolution of the 20,000 species of orchids: the plant is "infected" by the fungus soon after germination, before the root system has developed. The germinating seedling is nourished by the fungal partner, which, in orchids that are not photosynthetic, even supplies the plant with carbon compounds.

During the early stages of their evolution, land plants were geographically limited to low marshes beside lakes and watercourses. While water-based plants could disperse their spores through current and diffusion, the earliest land-based plants suffered the disadvantage that the offspring would have accumulated very close to their parents, competing for scarce resources. But 400 million years ago, in the Devonian period, as the clubmosses, giant horsetails, and ferns were emerging, another great innovation liberated land plants from their marshy environment.

Peter Atsatt has no doubt that the evolution of the embryo was one of the most important strategies evolved by plants in their colonization of dry land. Embryos allowed plants not only to solve the problem of competition between parents and offspring for space and nutrition, but also to improve the fitness of offspring by providing a store of nourishment, conveniently wrapped in the form we know as seeds, before they were launched into a distant and hostile world. This opened up another wonderful chapter of evolution, as we see today in the great variety of strategies for the dispersal of seeds. Embryos compacted to the size and lightness of dandelion seeds are carried on parachutes by the slightest breeze; the fruits of the coconut palm are designed to float to new beaches on tropical oceans; others have burrs and hooks that stick to the fur of animals that carry them to promising new destinations. The seeds of the fig, seductively packaged in sweet pulpy fruits, are a bonanza for many varieties of South American monkeys, including howlers and spiders, and for toucans, which throw the fruits into the air and catch them at the back of their throat; the seeds pass quickly through their digestive tracts and are deposited some distance away from the parent fig's canopy and root, individually rewrapped in a coating of compost.

While fruits are designed to be eaten, the seeds within are protected in a variety of ways, including armored casings, sheer numbers, and even poisons. Strychnine, one of the deadliest poisons known, protects the seeds of an evergreen that grows in the tropical forests of Asia. Squirrels and hornbills feed on the fruit pulp, but they take care not to crack open the disk-shaped seeds.

The success of this evolutionary creativity can be judged by the fact that plants forming embryos have spread so extensively that today the biomass of life on land is hundreds, perhaps thousands, of times that in the seas — and 84 percent of the mass of life on land is trees.

Flowers, one of the most aesthetically pleasing aspects of nature, give us great pleasure in the garden and home as well as through their associations with love and courtship, as the sonnets of Donne and Shakespeare attest. In more prosaic biological terms, they are the reproductive organs of a superstar phylum within the plant kingdom, the angiosperms, which runs to about 250,000 living species, compared to just 50,000 species of all nonflowering green plants.

The term "angiosperm" means a seed enclosed within a vessel. Their evolutionary origins in the Cretaceous, about 130 million years ago, puzzled Darwin, who described their sudden and inexplicable appearance as "that abominable mystery." Once arrived, however, the angiosperms displaced the highly successful gymnosperms, or "naked seeds," which include conifers, cycads, and ginkgos, to become the most successful group of plants that have ever evolved.

According to the Darwinian explanation, flowering plants arose from the preceding gymnosperms through a series of mutations, leading to a progressive rearrangement and transformation of their anatomy, with pollinating insects accorded an interactive role. But Kris Pirozynski, an emeritus paleomycologist at the Canadian Museum of Nature, in Ottawa, has speculated that flowers and fruits may be the results of symbioses involving fungi and gall-inducing insects. Pointing out that pollination by insects long preceded the arrival of flowers, he postulates that a carnivorous insect that laid its eggs inside the bodies of other insects may have altered its lifestyle to become a predator that laid its eggs inside the reproductive parts of plants, thus establishing a novel evolution. The eggs provoked a reaction from the plant that led to gall-like structures similar to fleshy fruit and seeds. Over time, and through a series of endosymbioses, the gall genomes were taken into the genetic code of the plant, leading to true fleshy fruits, seeds, and the great diversity and elegance of flowers we see today.

As with Atsatt's hypothesis of a plant-fungi symbiosis in the evolution of seeds in angiosperms, so Pirozynski's hypothesis about flowers and fruits must compete for credibility against the Darwinian explanation.

Such is the wonder of the new genomic age of biology that a modest plant only a few inches high known as thale cress is likely to shed light on both these hypotheses. In December 2000, *Arabidopsis thaliana* became the first plant to have its genome mapped, opening up potential new avenues not only in the field of botany but also in feeding the world. Whatever light the study of the plant's 25,000 genes will shed, on one fundamental point there is no disagreement.

The evolution of land animals was dependent upon, and therefore could only follow, that of land plants. This is such an important point that until recently it was believed that all the animals that have ever lived were dependent, either directly or indirectly, on plants for their day-to-day nutrition and, thus, for their evolutionary origins and survival. But recent discoveries have revealed an additional and equally astonishing story.

· 16 ·

THE CREATURES OF LAND, AIR, AND SEA

> The immense diversity of the insects and flowering plants combined is no accident. The two empires are united by intricate symbioses.
>
> — EDWARD O. WILSON, *The Diversity of Life*

TWO THIRDS of the Earth is covered by oceans a mile or more deep. Until the 1970s, biologists had assumed that these regions were lifeless deserts. The assumption was a natural one: sunlight penetrates only the surface water; photosynthesis stops at 100 meters, and virtually no light penetrates below 1,000 meters. But deep-sea fishermen had trawled up strange creatures from the abyss, including fish with enormous eyes, hinting at mysteries that remained to be solved. But before biologists could explore the deep oceans, engineers had to overcome the difficulties of constructing a submersible that would survive the immense pressures that build up at such depths. No advance hints could have prepared the scientists for the surprises in store. In 1977 a team headed by John B. Corliss, from the School of Oceanography, at Oregon State University, made a series of dives in the submersible *Alvin* onto the Galápagos Rift in the Pacific Ocean, midway between the islands made famous by Darwin and the east coast of South America. The rift, 1.6 miles deep, was known to be a zone where continental plates are pulling apart and new rock is being laid down as molten magma erupts onto the surface. The purpose of the *Alvin* exercise was to demonstrate the presence of thermal springs where convective currents through rock affected by this volcanic activity actively recirculated seawater.

En route to the abyss, Corliss and his colleagues from Oregon State, MIT, and the Woods Hole Oceanographic Institution repeatedly passed through a twilight world of fierce predatorial activity involving semitransparent creatures that had evolved ingenious strategies for survival. If the scientists flashed a beacon of light into the gloom, a mysterious phantasmagoria twinkled back in response. It was as if they had stumbled on a fabulous Milky Way of stars stretching away into the deeps. Each individual star represented a symbiotic union between a marine animal and one of the strains of bacteria that had evolved the ability to manufacture light through chemical pathways, a phenomenon known as bioluminescence.

Bioluminescence is the explanation for such eerie phenomena as fireflies, glowworms, and will-o'-the-wisps, which startle the eye by night on land. Bioluminescence had been observed before in marine life forms; for example, in the warm coastal waters of Southeast Asia, silvery shoals of fish known as leiognathids use its eerie light as a guide in mating and hunting, and even as a camouflage against predators. The chemical luminescence, controlled and directed by a lenslike device employing a complex arrangement of muscles, is the symbiotic contribution of billions of *Vibrio fischeri,* bacteria that live within the tissues of the fish.

In the deep ocean, the researchers discovered a range of adaptations on a scale that had never even been imagined before. Firefly squid signaled to each other with luminescent tentacles. Deep-sea anglerfish held aloft glowing lures that writhed and twisted like living morsels to attract the larger prey the angler had in mind. Others flashed to signal mates or switched on batteries of camouflage to deceive those that preyed upon them. Jellyfish wheeled and pumped like rapidly folding and unfolding umbrellas, their flesh bedecked with kaleidoscopic carousels of light.

At the very heart of the abyss was the strangest world of all. Here, where a series of volcanic summits spewed molten lava, the oceanographic scientists gazed on a sight that took them back billions of years to the Archaean phase of planetary evolution. Around giant natural chimneys that belched out black smoky water full of hydrogen sulfide, an unknown ecosystem opened up before their eyes. In the words of one of the scientists, "We discovered extraordinary communities of organisms living in the thermal vent areas at the rift axis."

Huge mussels of an unknown species proliferated over the floor near the vents. White crabs scuttled around cylindrical objects two meters high, from which strange carmine-red plumes expanded and retracted as pallid eel-shaped fish tried to eat them. The cylindrical objects were the "respiratory plumes" of an unknown genus of giant worm, now called *Riftia pachyptila.* In time biologists would identify more than five hundred new species of life in and around the vents. How was it possible for such species to evolve and survive in this lightless habitat, with very little dissolved oxygen? The adult worms were found to have neither mouths nor guts to eat and digest food. Their only nutritional source of energy came from symbiotic bacteria that metabolized the hydrogen sulfide in the geothermal fluids that bubbled out of the "black smokers." To date, microbiologists have not been able to culture the bacteria outside the worms' tissues, suggesting that the bacteria are as dependent as their hosts on the symbiotic relationship.

The red color of the respiratory plumes was identified as a highly specialized form of hemoglobin, the pigment found in blood. While in humans it is the main component of the red cells ferrying oxygen throughout the tissues, in the worms its free-floating molecules bind to both hydrogen sulfide and the scant supplies of oxygen available at such depths, ferrying the oxygen to the mitochondria and both oxygen and sulfur to the symbiotic bacteria. Vast numbers of these bacteria, contained in special organs deep within the worm's tissues, oxidize the highly poisonous hydrogen sulfide to provide the energy needed by the worm. This astonishing partnership of worm and bacteria was found to be more efficient at energy conversion than any sunlight-capturing photosynthesizer on land or in the oceans.

These findings provide an important clue to the origins of ancient life. The marine environment of the early Earth was rich in sulfides, and some evolutionary biologists are now convinced that the ability to breathe sulfur preceded the evolution of our oxygen-breathing eukaryotic ancestors.

The discovery of these creatures deep in the oceans inspired marine biologists to investigate anew many familiar life cycles in sulfur-rich environments and to revise many earlier assumptions. A range of sulfur-breathing symbioses is now being attributed to mussels, clams, and other mollusks that live around sewage outfalls and sea-grass beds. Most spectacular of all, in the depths of a cave in Rumania, the geologist Cristi Lascu and the biologist Serban Sarbu, of the Research Institute in Bucharest, discovered a unique ecosystem based on a mysterious variety of surface-floating bacterial mats. The food web included thirty-three previously unknown species of snails, wood lice, and eyeless shrimps, which fed on the bacteria and were themselves prey to leeches, centipedes, eyeless spiders, and water scorpions. The whole ecosystem was dependent on the oxidation of sulfur by this previously unknown bacterium.

These examples illustrate a stage of life that scientists now believe was the first step toward a food-web ecology. But there is an inherent flaw in any such ecology: the supplies of sulfur are finite. Even the deep-sea vents close down after a few decades, and the ecology has to begin again at some new site. Photosynthesis, first in water and subsequently on land, solved the problem of a finite energy source. In sunlight, life had a guaranteed and virtually limitless source of the energy needed for the metabolic machinery of all subsequent evolution.

Victorian geologists thought that life began 570 million years ago with the Cambrian period, named after the Welsh rocks that contained the first obvious animal fossils. But these were not the first animals to evolve; rather, they were the first to evolve hard body parts, such as shells, which could become fossilized after their soft parts had rotted away. Today we know that bacterial life existed long before that time. In fact, the forms that many biologists accept as the first true animals predate the Cambrian by at least 50 million years, a date that is likely to stretch back further as more discoveries are made.

Those first animals are known as the Ediacara, after the location near Sydney, Australia, where their fossils were first discovered. These animals had no internal or external skeleton, but their soft bodies left intaglio impressions in sandstone. Some appear to be the earliest annelid worms, others perhaps the ancestors of anemones and jellyfish or the forerunners of animals with jointed limbs, such as spiders and crabs or starfish and sea urchins. Others were strange creatures that subsequently became extinct. Even these were much more complex than the first real multicellular creatures, which are thought to have evolved more than a billion years ago. It is enlightening to realize that the Ediacarans evolved in the oceans before the first true plants. Unlike the terrestrial animals that followed, these creatures could not have dined on photosynthetic vegetation. Where did they obtain their food?

Many of these strange creatures apparently had no mouths or digestive tracts, which suggested to Mark and Dianna McMenamin that, like the *Riftia* living around the black smokers, they were dependent on bacterial symbionts for nutrition. An intriguing philosophy of cooperation is evoked by this earliest style of complex life: "Self-sufficient feeding strategies meant that nobody had to eat anybody else." The McMenamins' theory affords us a lovely vision of this strange Garden of Eden, with multicellular life first evolving as a peaceful "symbiotically driven" coexistence.

It is tempting to reflect on this vision of a primeval Eden. But one is inevitably led to ask: what Genomic step made possible the coexistence within a single organism of many different types of cells and the division of labor we now recognize as a cardinal feature of the tissues of plants and animals?

Smith and Szathmáry see life as evolving in five major transitions, in three of which symbiosis played a major role: the origin of the first cells (bacteria) from self-replicating entities, the origin of chromosomes to regulate these cells, and the origin of eukaryotic cells. In the fourth great transition, protoctists evolved to more complex, many-celled organisms in which groups of cells became specialized in ways that would, over time, give rise to the tissue-and-organ organization of an oak tree or a whale. In Smith and Szathmáry's view, the same "organizing" principles would give rise not only to fungi, plants, and animals but also to societies.

All animals are diploid (having two sets of chromosomes), multicellular, heterotrophic (dependent on other life forms for nutrition), and (except in certain cases, such as sponges) go through an embryological stage in which the animal consists of a ball of identical cells known as a blastula. In our human species, for example, the divisions of the blastula ultimately give rise to 100,000 billion cells, each adapting in some way to fulfill its specialized role in whatever tissue or organ it is part of; and all of the cells, tissues, and organs come together to make up a single coordinated human being. The evolution of such complexity was possible only with an equally spectacular evolution within the governing mechanisms of the Genome.

Darwin's supporter August Weismann was the first to consider the implication of this, and he concluded that cells could specialize into tissues in one of two ways. Translated into modern terms, either a different set of genes was passed to each tissue during embryological development or the cells in every tissue had to have identical genes. The latter explanation assumed that the Genome had discovered a way of controlling groups of genes, so that in each tissue or organ only those genes needed were "active."

Today we know that the second explanation is the correct one. Every cell in our bodies, except for the germ cells after reduction division, has an identical genome of forty-six chromosomes. The reason the transparent cells in the corneas of our eyes are so different from those making up the lobules of the liver, the muscle of the heart, or the nerve cells that send enormously attenuated axons down to the tip of the great toe is that in every specialized cell different sets of key genes are active while others are suppressed. The French scientists François Jacob and Jacques Monod made this important discovery.

Regulatory genes, which have the ability to switch other genes, or groups of genes, on and off, are themselves told what to do by "inducers" from outside the cell. In a human being, the mechanisms are very complex, with many different inducers, both positive and negative, influencing the system. The more deeply one examines this organization, the more labyrinthine it appears. For example, it implies a "dual heredity" system, in which the daughter cell receives not only a copy of the maternal DNA but also a copy of all the genes in their state of on-off activation. Key regulatory genes known as "Hox genes" are distributed at intervals throughout the chromosomes and act as directors in the embryological development of heads, eyes, antennae, limbs, and other body parts. Hox genes have an ancient pedigree: the same series of genes exists in worms, mollusks, the fruit fly *Drosophila* (in which they were first discovered), and humanity. The same gene that switches on the formation of a human arm might, in a fly, switch on the formation of a wing.

Some biologists see the Ediacarans as an interlude during which predators lagged behind the relatively large symbiotic grazers. Certainly with the arrival of the Cambrian, predators were making their mark on the evolutionary record. One of the most striking examples was *Anomalocaris,* which, worldwide in distribution, dwarfed all its fellow creatures with its six-foot body; it probably swam like a manta ray and, equipped with two large eyes on stalks, it searched for living prey, which it captured with two large spiked appendages. These passed the prey on to be devoured by a hideous pineapple ring–shaped mouth surrounded by multiple rows of fanglike teeth.

For the McMenamins, such large and brainy predators, able to seek out and capture mobile prey, put all of life under intense selection pressure to evolve protection. The fossil record confirms that early animals such as the trilobites had evolved the ability to make chitin, an important structural compound similar to the material of our fingernails and toenails, which in combination with calcium carbonate fashioned a strong, protective shell around their soft bodies. Further developments led to the exoskeletons of crustaceans, spiders, and insects. Much later, an even more subtle evolution within the soft bodies of primitive fish led to the inner skeleton, largely cartilage at first (as in present-day sharks), but evolving to the sculptural properties of bone. The vertebrates became the most successful of all the large-sized animal groups, the internal skeleton allowing for the supple flexibility of the cheetah, for example, which can arch its back to allow its hind limbs to cross its forelimbs and thus reach a maximum speed of sixty miles an hour in just a few strides. More important, the internal skeleton allowed vertebrates to grow without the need to shed their skins, shells, or carapaces. Think of the colossal dignity of the blue whale, the largest creature that has ever lived on Earth, measuring as much as 100 feet from tip to tail and weighing more than 200 tons.

But the vertebrates were latecomers in evolutionary terms. The animals that first radiated out over the land, the arthropods, which gave rise to today's insects and spiders, were far more humble.

The first of these included scorpions and flightless insects, which appeared about 410 million years ago. After the microbes, the insects became the most successful of all creatures, and by far the most populous animals even today, spreading and diversifying into every ecology, including the soil. They discovered flight and thus were able to colonize the air long before the birds, and in a few cases they recolonized water.

The evolving insects ultimately depended on photosynthesizing plants for their nutritional needs. While some, such as the sapsuckers, developed mouthparts to take nutrition directly from the vascular channels, a much larger variety simply ate the leaves and stems of living and dead plant tissues. But to derive nutrition from these sources they had to break down and digest the cellulose that made up the cell walls of the plants. As in the earlier struggle by plants to colonize land, symbiosis played a vital role in helping insects digest cellulose.

Experiments performed by Angela Douglas and others emphasize the importance of symbiosis when conditions are tough: many symbioses arise from the need to survive where the supply of nutrients is poor. Today, 70,000 species of insects depend on the bacterial and protist symbionts in their guts to help in the digestion of nutrient-poor diets. In insects that feed upon low-nitrogen plant material such as wood, the symbionts provide essential amino acids, while others fix nitrogen or recycle waste. The symbionts of insects that feed on plant polysaccharides break down and predigest the tough cellulose and fiber to a form that can be used by the host. Recent research on

Wolbachia bacteria, which are inherited by the host through the cytoplasm of the ovum, show that the symbionts interact at such a profound level that the relationship may produce evolutionary change, through manipulation of the host's gender, in as many as 16 percent of all known insect species.

Nowhere are the symbiotic cycles involving insects more aesthetically pleasing than in their evolutionary partnerships with flowering plants. The immense diversity of mutualisms involving insects and plants is no accident. The two empires are united by an amazing variety of symbioses at every level, including cohabitation, pollination, consumption of each other, and even cycling the products of each other's decay. The importance of these symbiotic cycles goes far beyond the individual life cycle. In his widely acclaimed book *The Diversity of Life,* Edward O. Wilson, Pellegrino University Professor Emeritus and curator in entomology at the Museum of Comparative Zoology at Harvard, states: "So important are insects and other land-dwelling arthropods that if all were to disappear, humanity probably could not last more than a few months."

The time needed for new life forms to evolve and adapt in the early days of life on Earth can be judged by the fact that 80 million years passed between the colonization of land by vascular plants and the time when an air-breathing fish first crawled out of the water to give rise to the amphibians. Symbiosis was vital in the subsequent evolution of all terrestrial animals. Every herbivore, including plant-eating lizards, birds, marsupials, and a wide variety of mammals, such as cattle, hippopotami, and giraffes, as well as the gigantic diplodocus and the triple-horned triceratops, could have evolved only in symbiotic partnership with its gut-based internal zoo of cellulose-degrading microbes. And as we now know, the symbiotic interactions went much deeper than this. In the leaves and grasses that became their diet, the energy of sunlight was trapped by endosymbiotic chloroplasts, which supplied the endosymbiotic mitochondria with the energy needed to metabolize oxygen. In the soil around the roots of virtually every plant, symbiotic fungi supplied the plant with needed water and minerals. In turn, the herbivores provided the nutrition for the evolving carnivores, from *Tyrannosaurus rex* to lions, tigers, and eventually the omnivorous ancestors of *Homo sapiens,* thus expanding the diversity of higher animal life.

Half a billion years ago, as the Ediacaran stage gave way to the early Cambrian, marine animals exploded into a period of wild evolutionary experimentation in the mud flats of long-lost oceans. The famous early Cambrian fossils in what is known as the Burgess shale were discovered by the geologist Charles Walcott in a mountainous bed of rock in British Columbia in 1910. Dynamited in huge numbers out of the shale, these fossil snapshots of evolutionary trial and error produced the most exotic monstrosities ever found, including *Anomalocaris.* Another animal, known as *Hallucinogenia,* takes the form of a sausagelike tube about an inch long, supported on seven pairs of pointed struts, with a bulbous drumstick (possibly the head) at one end, various wormlike appendages sprouting out of its back, and what one has to presume is an anus (but it could just as well be the real mouth) at the opposite end from the drumstick, looking for all the world like the spout of a kettle. While many of these forms do not appear to fit with any modern groups, others may well belong to one of the thirty-seven phyla that make up the modern animal kingdom, although not a single one of the Burgess shale creatures would be recognizable as a familiar species today.

However, by 400,000 years ago, we find fossils of a species that has endured and has found a niche in popular biology: the helmet-shaped *Limulus,* better known as the horseshoe crab, though it is not really a crab. Like the crabs, it is a member of the large phylum Arthropoda, which includes sea spiders, spiders, scorpions, mites, ticks, chiggers, and harvestmen. The eyes of *Limulus* are so primitive they probably cannot register a picture but only a sense of movement. In watching *Limulus* trundle slowly through the shallow waters of Cape Cod Bay for nocturnal spawning — the females excavate a depression in which they lay about three hundred eggs, which are fertilized by the smaller males, clasped to their backs with specially adapted pedipalps — we are privileged to witness a ritual that has likely changed very little in almost half a billion years.

An equally ancient animal is the coelacanth, which had been assumed extinct until it was found, just before Christmas 1938, by Marjorie Courtenay-Latimer, the curator of a small South African museum, and after whom it was named, *Latimeria chalumnae.*

Animals faced obstacles in their invasion of land, just as the plants had: the need to develop strong skeletal support to withstand gravity, to devise new techniques of feeding, mating, and respiring in air, and to guard against the dangers of desiccation. Of all the many forms of animal life in the oceans, very few managed to make it onto dry land. These included some arthropods that became the insects and spiders, an occasional mollusk, ancestral to the land snails, some annelid worms, and a few representatives of the group known as chordates because they possess a spinal cord. Like all the mammals living today, we belong to this group, and more specifically to a subgroup known as the vertebrates.

Further fossil discoveries, and growing understanding of the animal genome, will offer more details of the evolution of all land animals.

While the importance of Darwinian mechanisms in this great adaptive radiation should not be downplayed, the symbiotic contributions were not transient or temporary but enduring. In many cases, such as the mitochondria, mycorrhizae, chloroplasts, and the endosymbiotic sulfur-breathing bacteria, symbiosis underpinned all of the subsequent evolution of plants and the animals, becoming as permanent as those great divisions of life itself.

On August 14, 1970, in an article in *Science,* the biologist Peter Raven, then based in Stanford, highlighted the importance of endosymbiotic cellular evolution. Acknowledging that the evidence was now overwhelming that chloroplasts and mitochondria had arisen by endosymbiotic union of bacteria and primitive eukaryotic cells, Raven drew attention to the fact that this important new evidence was being ignored in all the current classifications. His concern was answered by Lynn Margulis, who, working with her colleague Karlene V. Schwartz, took SET to its logical conclusion and in doing so challenged prevailing thinking about the tree of life.

· 17 ·

A NEW TREE OF EVOLUTION

Even though natural selection is the evolutionary process that explains functional organizations, it is important to recognize that natural selection is not the only force in evolution.

— Elliott Sober and David Sloan Wilson, *Unto Others*

Although there had been many previous attempts to classify life forms, dating back to Aristotle and Pliny, the most practical classification was produced by the botanist Carolus Linnaeus, working in eighteenth-century Uppsala, Sweden. He gave every organism two names derived from Greek or Latin and conventionally italicized; the first name, always capitalized, denotes the genus, the second the species. For example, humanity is classified as *Homo sapiens,* meaning we belong to the human genus and to the intelligent species. Other human species have existed, but they are all long extinct, the most famous example being *Homo neanderthalensis,* Neandertal man. Whereas Neandertal man is not thought (by most but not all authorities) to have played a direct part in our own evolution, experts consider *Homo erectus* and *Homo habilis* our direct evolutionary ancestors.

Linnaeus grouped the genera into higher collections, or taxa, from which we derive the term "taxonomy," meaning the classification of life into an ordered system. The most fundamental grouping from an evolutionary perspective is the phylum. Each phylum embraces many subgroups, all of which began with a common major evolutionary step. For example, we humans belong to the phylum Craniata, or "craniates," which derives its name from the Greek *kranion,* denoting that what the members of this phylum have in common is a skull. In Linnaeus's day, all of the phyla were fitted into two great divisions known as the kingdoms of animals and plants. With the exception of Ernst Haeckel's modification, described below, this remained the orthodox basis of classification of the progressively branching tree of life. Symbiosis threatens this elegant sense of order. Through endosymbiosis, two or more species blend genomes in an evolutionary union, the opposite of classical Linnaean thinking. Widely separated branches of the tree come back together to create entirely new species — sometimes a whole new phylum or even a whole new kingdom.

By the late nineteenth century, a new look at the Linnaean system was timely. With the growing understanding of biology, life forms such as microbes had proved exceedingly difficult to classify. Viruses were the most intractable of all, considered by some to be living yet dismissed by others as curious entities somewhere on the threshold between life and nonlife. Botanists and zoologists frequently disagreed at the level of microbes, placing those that especially interested them in the kingdom they studied. For example, bacteria that moved were the preserve of the zoologists, whereas those that looked green, whether or not they moved, belonged to the botanists.

This prompted the German naturalist Ernst Haeckel to make some changes to the Linnaean system, placing some of the smallest life forms, neither plants nor animals, in a new kingdom he called the Monera. Haeckel's changes did not worry most biologists, who considered microbes the domain of pathologists and bacteriologists. The mainstream scientists were free to focus on their real interests, the so-called higher animals and plants.

In the last century, however, our realization of the huge diversity of microbes, coupled with a growing understanding of their role in evolution, has put previous taxonomic assumptions under strain. In 1956 Herbert F. Copeland, while accepting Haeckel's kingdom of the Monera to designate all of the bacteria, proposed a fourth kingdom, the Protoctista, to cover microbes that did not easily fit within the animal or plant kingdom yet had nuclei that distinguished them from the Monera. I have already mentioned that Copeland's book influenced Margulis when she first formulated SET. Whereas the bacteria were relatively easy to assign, the protoctists were much more varied and hence more challenging. Copeland decided to include many curious and little-understood groups in the Protoctista but, meaningfully in Margulis's opinion, he took the fungi, including yeasts, molds and mushrooms, out of the plant kingdom and added them to the protoctists.

As often happens with pioneers, especially those who intrude upon a scientific field from the margins, Copeland's four-kingdom classification was ignored at first. But a few years after it was published, it was adopted by Robert H. Whittaker, a professor at Cornell University who had founded the study of community ecology. Whittaker, like Copeland, realized that the Monera differed radically from the other kingdoms in having no nuclei. These prokaryotes were the evolutionary antecedents of the eukaryotes, organisms that have nucleated cells. But having studied mycorrhizal fungi in detail, Whittaker concluded that fungi, which could not photosynthesize, were so radically different from plants and from the protoctists that they also warranted a kingdom of their own. This led him to expand Copeland's four kingdoms to five.

Some years later, frustrated by the inconsistencies and contradictions that bedeviled the entire system, Lynn Margulis and her coworker Karlene Schwartz were searching for a more lucid approach that would help them teach their students "one consistent, comprehensive taxonomy that made sense." The five-kingdom concept, with some modification, fitted Margulis's vision. The tree of life was no longer seen as a simple stem with branches and bifurcating twigs but as "a twisted, tangled, pulsing entity with roots and branches meeting underground and in midair to form eccentric new fruits and hybrids." Margulis and Schwartz tackled the prevailing confusion by revising Whittaker and Copeland's views to include the evolutionary contributions of symbiosis.

Modern taxonomists have the advantage of electron microscopy, which can probe the tiniest organelles and structures within the cell, and the sharpest tool of all, the genomic probes of molecular biology. When the infrastructure of life is dissected at this level of precision, nonbiologists may be surprised to discover that an oak tree has more in common with a human being than either has with, say, a fungus, an amoeba, or a bacterium. For Margulis, far too much attention has been devoted to the kingdoms of plants and animals, and to the human species in particular: from the evolutionary perspective, a more extensive and revealing biodiversity is to be found within the new kingdoms of Monera and Protoctista.

For more than twenty years, Schwartz and Margulis have adopted the Whittaker system, collating a great deal of new information into its structure, meanwhile taking due note of the additional taxonomic implications of symbiosis. A single example illustrates the change in perception and the enlightenment it brings.

Early biologists, using simple staining and light microscopes, labeled tiny life forms they saw in water or mud as Protozoa, from the Greek words *proto* and *zoa,* meaning "first animals." In fact, these nucleated microbes have many characteristics that distinguish them from true animals: they are ancestral not only to animals but also to plants, fungi, and protoctists. This new classification does, however, leave room for some residual confusion. While some biologists use the term "protist" as synonymous with the protoctists, Margulis and Schwartz restrict this term to the simplest unicellular protoctists. The umbrella designation of Protoctista includes protists together with larger, often colonial forms, such as slime molds and seaweeds. This strange and little-known kingdom embraces an amazing 250,000 species. A great many that once existed are now extinct, although their microfossils are so abundant in the geological record that they are often used to date ancient deposits.

It would be satisfying to conclude that this is the end of the classification story, but science never works quite like that. Taxonomy is still controversial, and alternative proposals have been made by other authorities. Carl R. Woese, the innovative microbiologist at the University of Illinois at Urbana, has proposed a three-"domain" classification, with the most ancient life forms designated as a domain of their own, the "Archaea," which Woese considers equal in taxonomic importance to the Eubacteria or "true" bacteria, and the Eukarya, or all nucleated cellular life. This classification is based on significant differences in his molecular typing of ribosomal RNA. Woese's system has its adherents, but not every biologist is prepared to consider all of eukaryotic life as being only one third of the taxonomic system.

In an important respect, however, the two classifications reflect differences in philosophical perspective as much as biological perception. While Margulis is acutely aware of the whole organism, Woese takes a reductionist viewpoint focused on the molecular biology of the genome. From a neutral perspective, the two classifications have more commonalities than differences. It is significant, moreover, that both systems exclude viruses as living entities, compelling virologists to produce their own evolutionary classification for what, in my opinion, amounts to a sixth great kingdom of life. Biologists can choose the classification they find most accurate and useful, or they can sit around a table and look at the obvious commonalities and work out a compromise.

The taxonomic classifications are not just intellectual games for scientists to play at. Viruses apart, adopting either classification brings new insights into life. But even this wholly biological level of understanding is not enough to encompass the full wonder of evolution.

· 18 ·

THE CYCLES OF LIFE

There is a grandeur in this view of life, with its several powers, having been originally breathed into a few forms or into one; and that, whilst this planet has gone cycling on according to the fixed law of gravity, from so simple a beginning endless forms, most beautiful and most wonderful have been, and are being, evolved.

— CHARLES DARWIN, *The Origin of Species*

IN 1964 A CAMBRIDGE UNIVERSITY geologist, Brian Harland, found glacial debris, known as drop stones, close to the equator. This discovery made no sense because glaciers are found only at high altitudes and in polar regions. Harland was well aware that Earth has endured a series of ice ages, but none extended to equatorial regions. His observations led other scientists, including David Pollard and James Kasting of Pennsylvania State University, to the astonishing claim that about 600 million years ago the entire Earth froze into a gigantic snowball, with temperatures plummeting so low that glaciers really did form at sea level in the tropics. All of the great oceans were covered with thick sea ice, and the continents were blanketed in snow a kilometer deep. The populations of photosynthetic bacteria and algae were severely depleted, threatening the extinction of all surface life. Alternating periods of freezing and warming may have occurred at 10-million-year intervals as many as four times, threatening the extinction of life again and again.

Most geologists and biological scientists were skeptical of this theory, pointing out that because continents moved on the tectonic plates, surely the drop stones had been deposited when regions now at the equator were closer to the poles. But other scientists, in particular Joe Kirschvink, a geologist at the California Institute of Technology, showed that the magnetic fields of the drop stones confirmed that they had indeed been formed at the equator. Biologists were left to wonder how life could possibly have survived such a global freeze.

Nobody understood this conundrum until Kirschvink found part of the solution. Volcanoes protruding above the frozen surface would have continued to pump out carbon dioxide, the "greenhouse gas," which would have built up to such high concentrations in the atmosphere that it then brought about the most dramatic global warming since the Hadean period, when the Earth was still cooling from the fireball of its creation. But Kirschvink's scenario remained no more than theory until the early 1990s, when Paul Hoffman, a geologist at Harvard, and his geochemist colleague Daniel P. Schrag showed that layers of calcium carbonate rock laid down on top of glacial deposits in tropical areas such as Namibia could be explained as the result of high atmospheric carbon dioxide levels following the snowball period.

But if the entire Earth was frozen over, how did life survive? Chris McKay, a planetary scientist based at NASA, found the answer in icebound lakes of Antarctica's frozen valleys. Here, under ice 5 meters thick, he found an abundance of cyanobacteria and photosynthetic algae. How are they able to survive and proliferate? The answer, it seems, lies in the fact that when ice forms in such circumstances it is as clear as glass, allowing sunlight to penetrate in sufficient quantities that photosynthesis is still possible. McKay suggested that similar belts of relatively thin, transparent ice close to the equator would have provided a refuge for life during the millions of years when the seas were frozen over.

Earth's history has been characterized by long periods of relative homeostasis punctuated by major catastrophes. Of great interest are the extinctions that occurred after the arrival of multicellular life. The most familiar of these took place at the end of the Cretaceous period, some 65 million years ago, an extinction that may have been caused by climate changes following the impact of a giant asteroid or comet in the Gulf of Mexico opposite the Yucatán Peninsula. That extinction had great significance for the subsequent evolution of *Homo sapiens,* for in ending the reign of the dinosaurs, it opened the evolutionary door to the age of the mammals.

The most terrible of the extinctions, about 250 million years ago, wiped out 95 percent of the life forms that had struggled into existence over the 300 million years since Snowball Earth. In all, five catastrophes have wiped out more than half of the species existing at the time, and at least ten more have resulted in lesser extinctions. In all these disasters there is a salient message: each resulted in great disruptions of the biosphere, yet in every case equilibrium was eventually restored.

This ability of the Earth to rebound from catastrophe has caused some scientists to reexamine the validity and relevance of James Lovelock's hypothesis of an Earth in which life and the inanimate world interact in such a way that the planet is essentially self-regulating.

We obtain an exquisite level of understanding from the reductionist thinking of neo-Darwinism, as in Dawkins's detailed analysis of "the accumulation of small change" that gave rise to the almost perfect machinery of the ear and eye. But we learn just as much in exploring the fact that life and landscape, the makeup of the atmosphere, the chemical constituents of soil, rock, river, and ocean, are wholly interrelated. The difference in appearance between the Earth and Mars makes this clear. Neo-Darwinism does not rule out Gaian theory, any more than an understanding of the great forces at work in our biosphere diminishes the wonder of the creative Genome. Both levels of understanding are essential to appreciation of the true grandeur of life and its evolution.

The word "nature," from the Latin *natura,* has a number of definitions and interpretations. Amalgamating two of the commonest definitions from *Webster's College Dictionary,* we arrive at "the universe, with all its phenomena, together with the laws and principles that guide the universe or an individual." Integral to the laws and principles of nature are the great cycles of life. These remain valid and valuable for two reasons. First, except for the forces of nuclear fission and fusion, the elements that make up life — such as oxygen, nitrogen, and carbon — are indestructible, so they are continually recycled. And second, the quantities of the vital elements available for the processes of life are limited, so the oxygen you and I breathe at any moment has previously gone through the life cycles and bodies of a myriad other creatures. We now know that Earth's atmosphere is so thoroughly mixed and so rapidly recycled that, in the words of Preston Cloud and Aharon Gibor, "the next breath you inhale will contain atoms breathed by Jesus at Gethsemane and by Adolf Hitler at Munich."

We need consider only a few cycles to share the wonder that inspired Lovelock. Before the advent of life, Earth's atmosphere was very different from what we breathe today. Rich in carbon dioxide and with some nitrogen and traces of hydrogen sulfide and hydrogen, it was closer to the atmosphere we now detect on neighboring planets such as Venus or Mars. Iron and other elements saturated the oceans, but these could remain in solution only in the absence of oxygen. With the advent of photosynthesis, all of this changed. Today oxygen makes up about a quarter of the atoms in living matter, and virtually all the oxygen we breathe derives from photosynthesis. This made possible the further evolution of animals and plants, with their specialized cells, tissues, and internal organs.

However, an oxygen-breathing biosphere is an astonishing paradox. Oxygen is extremely reactive chemically, corroding iron and devastating a wide range of the metabolic reactions that are essential to life. Nevertheless, our human cells can obtain their high-energy requirements only by breaking down "fuel" chemicals, such as carbohydrates, in the presence of oxygen.

To illustrate what this means, take the energy a living cell can obtain from the fuel molecule we know as glucose. When it is broken down in the presence of oxygen, a mole of glucose (its molecular weight in grams) gives up 686 calories of energy; if it is broken down by the nonoxidative metabolism known as fermentation, it yields just 50 calories. Nature therefore selected the evolution of our red blood cells, which carry oxygen to every living cell. Yet every one of those cells needs protective mechanisms to prevent the corrosive oxygen from coming into contact with many of the components of its metabolism.

Oxygen, once released from water molecules by photosynthesis, accounts for 21 percent of the gaseous mixture that is our atmosphere, from which it enters many mechanisms and pathways. In the biological world, it is taken up for respiration by every plant, animal, fungus, and even the aerobic protoctists that have entered into symbiotic union with oxygen-breathing mitochondria. One variant of exosymbiosis is the sharing of the chemical products of metabolism, known as metabolites, by different life forms. Does it matter conceptually whether the metabolites are shared in a close physical relationship, or whether they are carried from one life form to another through the more diffuse medium of air or water? The evolutionary implications are the same in both cases. Oxygen is a metabolic product of photosynthesizing cyanobacteria, algae, and plants. When these life forms release it into the atmosphere and oceans to be used in respiration by other plants, protoctists, fungi, animals, and even aerobic bacteria, they are participating in a diffuse form of exosymbiosis I have labeled "macrosymbiosis," to indicate that it takes place at a planetary level.

The oxygen cycle is one of a series of such planetary gas cycles. Another is the carbon dioxide cycle. In Earth's present atmosphere, carbon dioxide is a trace gas, stable at 340 parts per million, which is less than one one-thousandth of the levels that existed before the arrival of life. Like oxygen, carbon dioxide interacts in important ways with both life and inanimate nature. It is a basic building block of the food webs, taken up by photosynthesizers and built into carbohydrates. It also plays an important role in the temperature regulation of the planet. Venus is overheated with greenhouse effects because it has 300,000 times Earth's level of atmospheric carbon dioxide, while even chilly Mars has 20 times Earth's level. Until the Industrial Revolution, with its contamination of the environment with greenhouse gases, carbon dioxide levels were progressively, if intermittently, declining. The decline was so great that 10 million years ago selection pressure led to a metabolic innovation in grasses; an advantageous mutation enabled grasses to photosynthesize more efficiently at low carbon dioxide levels.

In the 1970s geologists tried to explain the low levels of carbon dioxide in Earth's atmosphere using a strictly geochemical argument. Assuming that the sole source of carbon dioxide was volcanic emissions and the only removal system was its reaction with calcium silicate in rock, they assessed the effects of the warming sun and rainfall on the interaction between the carbon dioxide being poured into the atmosphere by the volcanoes and its removal by chemical combination with rocks. The result of their calculations was a system involving negative feedback, which kept down the level of carbon dioxide in the atmosphere. But the level predicted for today's atmosphere was between ten and one hundred times more than is actually observed — closer to that found on Mars.

When Lovelock added a Gaian mechanism of living interaction to the geologists' model, the results were a much better fit with today's carbon dioxide levels. As he explained to me: "I looked at what data there was on the levels of carbon dioxide between grains of soil in vegetated ground and, to my amazement, I found that the actual carbon dioxide level was thirty times higher than that in the atmosphere. I concluded that organisms in the soil were combusting organic matter, using up oxygen and carbon and producing high local levels of carbon dioxide. The rocks were exposed not to 340 parts per million of carbon dioxide, as the geochemists thought, but to anything up to 10,000 parts per million. Not only does this mean that the reaction of carbon dioxide on the rocks will be greatly increased, but the rate of increase is also temperature-dependent. If it is very cold, the contribution of these living organisms will be very little, but when it is hot they will produce a lot more carbon dioxide. So you've got a negative feedback mechanism in which life is playing an important part. High atmospheric carbon dioxide raises the temperature of the atmosphere and surface of the planet, in turn speeding up the metabolism of the life forms that remove it again by converting it into bicarbonate."

Without this carbon dioxide pump, that gas would increase in the atmosphere and the Earth would, over time, heat up. Lovelock takes the view that the falling levels of atmospheric carbon dioxide are, from a Gaian perspective, a necessary compensation for the fact that the sun has been getting hotter since life began. This contribution of Gaian theory to the carbon dioxide cycle was published in two scientific papers, causing geologists, atmospheric scientists and geochemists to reappraise their theories.

One of the most important applications of Gaian theory began in the early 1970s, at a time when Lovelock and his family were living in the rural retreat of Adrigole, a small town overlooking Bantry Bay in the far southwest of Ireland.

As he explained to me, "I used to walk along the seashore, which was no more than a few hundred yards from the cottage where we lived, and, being the odd sort of person I am, I took my gas chromatograph along with me and I tested the gases that came off the seaweed. I knew that there was a famous Professor Challenger at Leeds who, in the 1950s, had discovered that dimethyl sulfide came from a marine alga called *Polysiphonia fastigiata (P. lanosa)*. There happened to be plenty of this seaweed growing on the shore. Sure enough, when I tested it, I found that it gave off enormous quantities of this gas. I later confirmed this in other experiments. Anyway, that was where it all started."

What started on the Irish shore is a remarkable story, one that helped us understand not only the global threat of atmospheric pollution, acid rain, and the effects of chlorofluorocarbons (CFCs) on the ozone layer, but also illustrated the role of life in the global regulation of climate. After publishing two papers on atmospheric pollution, which provoked little interest in the scientific community at the time, Lovelock continued to mull over the significance of his findings. "What," he asked himself, "if these gases are emitted not only by the algae on the shore but all through the oceans?"

The natural cycle of sulfur had not yet been established. It was well known that the great bulk of sulfur on land was converted to sulfate ions, which washed into the rivers and then into the oceans. Without some mechanism of return, life on land would have been starved of this essential element. Geochemists believed that the oceans produced large quantities of hydrogen sulfide gas, which entered the atmosphere and was recycled back to land. But Lovelock was aware that hydrogen sulfide was rapidly oxidized in water containing dissolved oxygen, so very little of it would be likely to escape into the atmosphere. Moreover, Professor Challenger's work had pointed to dimethyl sulfide rather than hydrogen sulfide as the gas produced by marine organisms. "It seemed to me that, on both these counts, hydrogen sulfide could not be the major carrier of sulfur from the oceans to the land." Lovelock decided to look further into the sulfur cycle.

As so often in his previous attempts to pioneer new thinking, all of his applications for research funding were turned down. But he managed to inveigle his way aboard the research ship *Shackleton* on a voyage to Antarctica. He was given free board and lodging, but he received no salary and covered all the costs of his scientific work out of his own pocket. On the voyage he repeatedly measured the sulfur-carrying gases in the air and sea from the Northern to the far Southern Hemisphere. His measurements confirmed that not only was dimethyl sulfide present throughout the oceans but also that chlorofluorocarbons were persistent and long-lived in the Earth's atmosphere, a finding that would ultimately lead to the "ozone war." Yet another discovery — that two halocarbon gases, carbon tetrachloride and methyl iodide, were found in the atmosphere wherever the ship sailed — helped in our understanding of the chlorine and iodine cycles.

Peter Liss, an ocean chemist at the University of East Anglia, looked hard at Lovelock's data and found, to his astonishment, that although there are many industrial sources of sulfur pollution, almost all of the "natural" sulfur cycle is carried out through dimethyl sulfide. These findings were largely ignored until, in the early 1980s, the German geochemist M. O. Andreae confirmed that dimethyl sulfide was the major carrier of elemental sulfur from the oceans to the land.

In 1986 a fortuitous coming together of minds extended this discovery to the interactive role of life in the regulation of global temperature. That year the University of Washington in Seattle invited Lovelock to be a visiting professor. After giving his standard Gaia lecture to the chemistry department, Lovelock was approached by a meteorological chemist, Robert Charlson, who was baffled by the formation of clouds over the oceans.

The oceans cover two thirds of the planet, and the Pacific alone, which is largely devoid of landmasses, accounts for roughly half of the Earth's total surface area. The formation of clouds over such a vast expanse of water plays a major role in regulating climate. Charlson pointed out that clouds form only around small particles called "condensation nuclei." Water vapor does not simply rise out of the oceans and condense as droplets. If it did, the vapor would condense as marble-sized drops that would immediately plummet to the surface. To remain light enough to float in suspension, clouds need ultramicroscopic nuclei around which tiny droplets of water can condense. According to the standard texts, these nuclei were sea-salt crystals. But Charlson had flown over the Pacific Ocean searching for these crystals and had never found them where clouds were condensing. Many years later Lovelock would recall Charlson shaking his head and demanding to know, "Where the heck are these nuclei coming from?"

Lovelock suggested the sulfuric acid produced by factories in huge quantities as possible nuclei for cloud formation. But Charlson shook his head. Sulfuric acid particles fell out of the atmosphere before they traveled 600 miles. Clouds formed at distances much farther than 600 miles from any land-based industrial sources.

It suddenly dawned on Lovelock that his own experimental findings might provide the answer. He explained that algae throughout the oceans are constantly producing dimethyl sulfide, which enters the atmosphere and then disappears quite rapidly. Might it be turned into microscopic droplets of sulfuric acid, which in turn formed the nuclei of clouds? Charlson was very interested in this notion. But how could they check it? Putting their heads together, the two scientists worked out that if dimethyl sulfide was being converted to sulfuric acid, one of its oxidation by-products, methane sulfonic acid, which is not produced by any industrial source, would show up in the atmosphere. If they could demonstrate that this chemical was indeed a component of the atmosphere over the oceans, they would have their answer. When Lovelock and Charlson looked for methane sulfonic acid, they found it. Dimethyl sulfide, produced by algae in the oceans, was rising into the atmosphere, where it was oxidized to sulfuric acid, which in turn provided the condensation nuclei for cloud formation. The implications were staggering. Charlson calculated that clouds, in reflecting back the heat of the sun, cooled the global climate by about 18°F.

They published their findings in a paper coauthored with two other scientists, M. O. Andreae and Charlson's graduate student Steve Warren. Within a year the authors were awarded the Norbert Gerbier Prize of the World Meteorological Association.

After some time, and a good deal more research, the truth was found to be more complex than had been described in the initial paper. In the mid-1990s, Lovelock collaborated with a young American scientist, Lee Kump, to consider the fact that at temperatures above 53°F, the top layer of the ocean stratifies. Water from below cannot get to the surface and the heat kills off the surface algae. That is why tropical oceans are bright blue and clear while the Arctic Ocean is as turbid as soup. In Gaian thinking, the algae have evolved in such a way that they help themselves by seeding clouds that cool the surface layers of the oceans.

Gaian theory offers many other applications, such as the recycling of energy and even the recycling of that most basic commodity of all, the 1.5 billion cubic kilometers of water in the Earth's water cycle. Other cycles involve the circulation of gases such as nitrogen and methane, or elements such as phosphorus, calcium, magnesium, and silicon. The evaluation of these cycles, which is already under way, will involve research in many fields, including geology, meteorology, geochemistry, and biology, before we can understand the integral circuits and feedback loops involved in this gargantuan concept of Gaia.

After *Nature* turned down Lovelock's first Daisyworld paper, the editor, John Maddox, expressed his regret and suggested he submit the next relevant paper he wrote on Gaia. He did so, and the dimethyl sulfide explanation of clouds was the lead article.

Meanwhile the models of Daisyworld became more complex. The distinguished ecologist Robert May of Princeton University, in the book *Theoretical Ecology*, suggested that as a system becomes more complex, in the sense of additional species and a greater degree of interdependence, it also becomes more fragile. But extrapolations from Daisyworld modeling suggested that species would not grow uncontrollably, as had been suggested by the calculations of ecologists such as May: if they did, the environment became unstable and growth was curtailed. Lovelock tested this scenario with Daisyworld models of increasing complexity and interactiveness. He even introduced a series of environmental catastrophes. But the system remained stable, the daisies continued to regulate the climate, and their populations recovered.

Climatologists took this modeling further. Groups working in Britain and Germany used it to investigate the interaction between local climate and the great coniferous forests across Siberia and Canada. They concluded that the forests were having a huge climatic effect. They were behaving exactly like black daisies, absorbing sunlight and heating up their regions to a significant degree. By degrees, as ever more complex models of Daisyworld were constructed and as numerous applications were found to fit the observed facts in nature, important opinions swung round to supporting it.

The last to come on board were leading neo-Darwinians, notably William Hamilton of Oxford. As mentor to Richard Dawkins, he had played an important role in formulating the "selfish gene" approach. He had also been extremely hostile to Gaia. But after he realized that Lovelock had introduced the possibility of cheating into his Daisyworld modeling, Hamilton reconsidered his opposition. Lovelock had indeed included daisies that were a neutral gray color and had a 5 percent advantage over the others. According to Darwinian theory, they should have taken over. But Daisyworld showed that they did not. After much argument and debate on a number of occasions and over time, Hamilton finally remarked to Lovelock, "You have convinced me that the Earth is self-regulating, but I can't for the life of me see how natural selection could ever possibly have led to it."

Hamilton's conversion appeared complete in a paper he wrote with Lovelock's successor, Tim Lenton, that examined the importance for organisms of finding ways to spread their seeds. The title of the paper was "Spora and Gaia." One suggestion arising from Hamilton's new line of thinking was that through dimethyl sulfide, ocean-living algae may have evolved a mechanism of stimulating wind currents to spread their airborne spores to richer pastures. At this point Hamilton became very interested in Gaian theory, performing model experiments on the regulation of dams by beavers. He delivered a paper on this subject at a meeting in Oxford, declaring at the end of his talk that he now saw Gaia as a new Copernican revolution in biological thinking. "We wait," he said, "for an Isaac Newton to explain how it all works."

In a subsequent meeting, at which both Lovelock and Dawkins spoke, Dawkins, commenting on the drawn-out controversy between Gaians and neo-Darwinians, stated that it was largely a problem of the behavior of their disciples. "Jim has had his and I have had mine, and they have been far too enthusiastic and quite often got it all wrong." Tragically, soon after this meeting William Hamilton died unexpectedly from malaria contracted during a field trip to Africa. His death was a great loss in many ways, not least for the fact that of the leading proponents of neo-Darwinism he was the best equipped to continue to probe its interactions with Lovelock's theory.

Meanwhile, the broader world of environmental scientists was in no doubt as to the importance of Gaia. In July 2001 a meeting of the European and American Geophysical Unions took place in Amsterdam under the auspices of four globally based programs. At the conclusion, roughly a thousand scientists arrived at a declaration that, given the history of the Gaian controversy, must be considered a new enlightenment: "*The Earth System behaves as a single, self-regulating system comprised of physical, chemical, biological, and human components. The interactions and feedbacks between the component parts are complex and exhibit multi-scale temporal and spatial variability.*" This validation of Gaian theory had important implications for humanity and our exploitive behavior in relation to the biosphere. The declaration continued: *"The understanding of the natural dynamics of the Earth System has advanced greatly in recent years and provides a sound basis for evaluating the effects and consequences of human-driven change."*

At the time I interviewed Lovelock, in August 2001, he had finally, at age eighty-three, relinquished the reins of further experiment to younger scientists such as Lenton. His concluding words will come as no surprise: "The quest for Gaia has been a battle all the way."

He could have been speaking about evolutionary theory in general.

· 19 ·

THE SOUND AND THE FURY

I am very interested in the Universe — I am specializing in the universe and all that surrounds it.

— PETER COOK, *Beyond the Fringe*

ON DECEMBER 9, 1981, Stephen Jay Gould found himself fighting in defense of Darwinism in a U.S. district courthouse in Little Rock, Arkansas. He had been called as a witness in the key case of *McLean* v. *Arkansas Board of Education.* An act passed the year before by the Arkansas state legislature had required science educators to teach biological evolution from both an evolutionary perspective, based on science, and from the latest creationist point of view. There were good reasons why the plaintiffs had called Gould to give evidence. Creationists were suddenly proclaiming — albeit, from Gould's viewpoint, "with their usual skill in the art of phony rhetoric, cynically distorted" — that Gould himself had "virtually thrown in the towel" and admitted that the fossil record ran counter to the central tenet of Darwin's evolutionary theory.

Of course, from the first publication of Darwin's *Origin,* the history of evolution has been one of battles and intense rivalries, not only between evolutionists and creationists, but also between individuals within the field of evolutionary theory itself.

The great evolutionary geneticist Theodosius Dobzhansky engaged in battle with his Nobel Prize–winning rival, Hermann J. Muller. Most, if not all, of the major contributors to the "synthesis" breakthrough that founded neo-Darwinism bolstered their ideas with personal foibles and prejudices. Some disputes, for example that between the British mathematician Ronald A. Fisher and the American evolutionary biologist Sewall Wright, were every bit as rancorous as those between the selectionists and saltationists and between both of these persuasions and the creationists.

Even in the tighter scientific arenas of the twenty-first century, no scientist really works in isolation. In the words of the scientific philosopher Michael Ruse, colleagues are "always circling, wanting to grab the goodies: students, grants, research space, places in the curriculum." It should come as no surprise, therefore, that Lynn Margulis derides neo-Darwinism as a minor Anglo-Saxon obsession or that John Maynard Smith defends his patch with equal righteousness, bridling at Margulis's argument that symbiosis is the main source of evolutionary novelty. "This," he declares, "will not do. Symbiosis is not an alternative to natural selection: rather, we require a Darwinian explanation of symbiosis."

It was against such a background that the American paleontologists Stephen Jay Gould and Niles Eldredge first entered the fray. In 1972 they fretted over the apparent dichotomy between the fossil record and the central tenet of classical Darwinism, that natural selection occurred through the gradual accumulation of small changes. This theory had embroiled Darwin in disagreement with supporters such as Huxley, but Darwin had refused to budge: life, he insisted, does not indulge in saltations, but changes slowly and gradually over great periods of time. But even Darwin was forced to acknowledge that in the fossil record new species and even whole new phyla seem to appear out of the blue. Moreover, and to the great joy of the creationists, the emergent species and phyla have persisted with relatively minor changes over very long periods.

Darwin asked himself: "Why then is not every geological formation and every stratum full of such intermediate links? Geology assuredly does not reveal any such finely graduated organic chain; and this, perhaps, is the gravest objection that can be urged against my theory."

This puzzle, which might be termed the "fossil paradox," was swept under the carpet for more than a hundred years, attributed to the scantiness and patchy nature of the fossil record. But Darwin had also emphasized the importance of geographic isolation, for example in the Galápagos Islands, in the origin of species, and in 1958 Ernst Mayr built on Darwin's ideas, explaining that when a small group from a large and populous species became geographically isolated, new species would split away from the existing one. The formation of new species by splitting off from a previous one is called speciation, and Mayr called his theory "allopatric speciation," or the formation of a new species as a result of isolation.

Gould and Eldredge took this idea further in an attempt to explain the otherwise inexplicable gaps in the fossil record. Niles Eldredge set this controversial ball rolling in 1971, when he used Mayr's theory to explain the sudden changes that appeared in fossils of Paleozoic invertebrates. Eldredge's article so intrigued the science editor Tom Schopf that he sponsored a symposium at the 1971 meeting of the Geological Society of America and invited Eldredge and Gould to contribute. Gould was assigned to talk about speciation, a subject he professed to know nothing about. But when he asked for a different assignment, he was instructed to proceed or "get off the pot." Gould then teamed up with Eldredge, his classmate from graduate school at the American Museum of Natural History, on the lecture for the symposium, which later became a chapter in a book, also edited by Schopf.

Gould and Eldredge argued that the fossil record was not patchy or misleading but perfectly accurate and revealing. They proposed that mutation within a relatively small gene pool would give rise to a much quicker pace of evolution than occurs in a large population. Most mutations would make little or no difference to the broad sweep of evolution, but when a variation emerged that had a major selection advantage, it would subsequently spread far and wide. The isolated locality of origin and the short duration of change left little mark in the fossil record; the successful new species would appear suddenly in the fossil record and would survive in relatively large populations and for lengthy periods.

This model, which became the basis of a subtheory of Darwinian evolution known as "punctuated equilibrium," had implications for the level at which natural selection took place.

For example, within a major species, large in number and covering a wide geography, many new species will arise through a series of geographic isolations. Selection takes place not at the level of the individual, as classical Darwinian thinking held, but at the level of competing subsets of species; one subset is more likely to survive in a particular environment. Gould and Eldredge were content that they had explained the fossil record: what was being found in that record was exactly what would be expected and thus their theory paved the way to a new understanding of, and future research in, paleontology. Gould later confessed that he had anticipated nothing more than a modest reaction from colleagues to the punctuated equilibrium theory. Prior to publication, he showed the article to his father with the comment, "Nobody will read it, and no one will pay any attention." His father, however, disagreed: "This is terrific," he exclaimed. "It will really shake things up." And it did.

Anything that challenges Darwinism in the slightest manner is treated by neo-Darwinians as a threat to the gospel. Punctuated equilibrium provoked a "major brouhaha" among scientists and lay readers, enmeshing its authors in a simmering controversy that continues to the present day. It was the subject of editorials in the *Times* of India and articles in the Beijing *People's Daily*. Richard Dawkins dismissed it as unnecessary in a chapter of *The Blind Watchmaker.* Ernst Mayr made essentially the same point, though more briefly, in his book *One Long Argument.* Creationists, on the other hand, welcomed the new theory with open arms, delighted with the notion that the fossil record contradicted the core concept of Darwinian evolution.

So it was that Gould found himself in the witness box before Judge William R. Overton of Arkansas in the 1981 trial. In Gould's words, "Most of my testimony . . . centered upon the creationists' distortion of punctuated equilibrium." From an analysis of the trial published fourteen years later by the theology professor Langdon Gilkey, it is clear that the crucial decision was whether or not creationism constituted a valid scientific theory to compare and contrast with evolutionary theory. The "overwhelming weight of scientific evidence" proclaimed it did not, and Judge Overton's decision in favor of the evolutionists was upheld in June 1987 by the U.S. Supreme Court.

Five years after their first article was published, Eldredge and Gould, though somewhat battered and bruised, reissued their challenge to conventional wisdom: "We believe that punctuational change dominates the history of life; evolution is concentrated in very rapid events of speciation (geologically instantaneous, even if tolerably continuous in ecological time)." Gould has stated that many studies in paleontology have now confirmed that punctuation is a very real and important phenomenon, so much so that it has become "an ordinary instrument of active research."

It seems likely that Darwinian mechanisms have played an important role in many examples of speciation, as in the great variety of the mammals; within every genus, selection has led to the "adaptive radiation" of the component species. However, in considering sudden evolutionary change, one should include not only Darwinian mechanisms but also the capacity of symbiosis to produce very rapid change. The arguments of Dawkins and Mayr, and for that matter Eldredge and Gould, do not even consider symbiosis as a source of the fossil saltations, though David Bermudes and Richard C. Back, working at the University of Wisconsin's Center for Great Lakes Studies, draw attention to precisely this possibility. In analyzing the fossil record for evidence of symbiotic evolution, these authors point to the findings of Margulis and others that endosymbiosis has given rise to many new phyla, suggesting that this is "one possible mechanism for the discontinuous appearance of species described by Eldredge and Gould."

Indeed, the omission of symbiosis by Eldredge and Gould is bewildering, since even Smith and Szathmáry accept that symbiosis may have given rise to three of the five major transitions of life.

In her book *The Symbiotic Planet,* Lynn Margulis recalls a conversation with "the eloquent and personable Niles Eldredge." She asked him if he knew of any case in which the formation of a new species had been documented, whether in the laboratory or from observations taken in the field or from the fossil record. Eldredge could think only of a single example, from Dobzhansky's experiments with *Drosophila.* In one experiment, flies bred at different temperatures separated into two distinct populations, hot-bred and cold-bred, which could no longer interbreed. The inability to interbreed is a fundamental characteristic of speciation. Then Eldredge dismissed his own example with the recollection, "But that turned out to have something to do with a parasite." In fact the species change in the fruit fly resulted from the elimination of a temperature-sensitive symbiotic microbe.

In 1990, while working in Lederberg's laboratory at Rockefeller University, Jan Sapp was discussing whether or not sociopolitical forces had contributed to the downplaying of symbiosis. He remarked to Lederberg, "Well, the interesting thing is that symbiosis certainly challenges individualism, and in this light it seems relevant that there has never been a meeting on symbiosis in the United States." Lederberg immediately went back to his office to check on this assertion, and returned with the remark, "You're right. I can't believe it." The two scientists realized that the bulk of the journal literature on symbiosis had been European until Margulis began writing in the 1970s. In the United States, neo-Darwinism, with its rigid adherence to Mendelian genetics and the nuclear monopoly of heredity, reigned supreme. So it was easy to see why, during Margulis's discussion with the "personable" Eldredge, the concept and mechanisms of symbiosis were still largely misunderstood.

But the negative attitudes toward symbiosis are changing. From the mid-1970s onward, ecologists have begun to take more notice of symbiotic associations. By the 1980s Douglas H. Boucher at the University of Quebec had found thirty studies in which models that previously had focused exclusively on competition could be usefully reappraised from the point of view of symbioses. To turn a competition model into one of mutualism, all one had to do was to reverse a sign in the equations. Boucher gave examples from diverse ecosystems, including the midocean diatom mats, the sulfur-rich deep-sea trenches, trees and epiphytes, rice paddies, and a great many more. Similar evidence was enough to change Robert May's opinion in favor of mutualism, and in 1976 three separate studies showed how interactions that appeared to be predatory or competitive might actually be mutualisms. In that same year the *Bulletin of the Ecological Society of America* commented that "ecologists had neglected to look for mutualism and thus had failed to note its importance."

We have seen how endosymbiosis gave rise to at least twenty-eight of the seventy-four eukaryotic phyla. Bermudes and Back, taking this idea further, have shown how symbiosis might have contributed to the origins of several important marine fossil groups. Even if every case cited has not been proven, it is likely that a substantial proportion of the big evolutionary changes resulting in phyla and kingdoms arose from the endosymbiotic merging of genomes. Bermudes and Back make plain that this is a form of macroevolutionary saltation. As more critical examination takes place, symbiosis will be seen as having contributed to many more examples of evolutionary change at all taxonomic levels, from phyla to species.

Michael Dolan, an evolutionary biologist in the department of geosciences at the University of Massachusetts, is currently working on the many species that make up an order of "swimming protists" known as hypermastigotes. Most of the component species are involved in symbiotic relationships with bacteria. Some of these bacteria are long and thin, some ciliated and some not; others are round and short, or even a combination of the two. Rather than classifying the species in terms of conventional Linnaean differences in the protoctist, Dolan is collecting information that will assign them to species on the basis of their symbiotic partners.

Donald Williamson, a marine biologist at the Port Erin Marine Laboratory, on the Isle of Man, has proposed a saltationist theory for the origins of larvae that, if confirmed, would be among the most astonishing examples of sudden change in all of evolution.

The Roman writer Ovid wrote a poem based on myths from many ancient sources, which he gathered together under the title *Metamorphoses.* The physical transformation of one living form into another seems utterly bizarre even in the worlds of poetry or fantasy, yet it is the norm in the life cycles of a great variety of insects and marine animals. Some of these real-life metamorphoses are, like Ovid's stories, both enthralling and exquisitely beautiful to behold; for example, the unfolding of the fully formed butterfly from its pupation in the chrysalis. Others more closely resemble the science fiction film *Alien.* Conventional Darwinism assumes that the larva is a more primitive evolutionary stage of the adult, and, during metamorphosis, the larva merely changes into the adult. But in the starfish *Luidia sarsi,* the adult grows as a separate entity within the tissues of the larva, and when it is sufficiently mature, the adult tears through the outer skin of the larva and migrates to the ocean bed, leaving the swimming larva to carry on an independent existence. Many other marine organisms display anomalous features in their larval stages that cannot easily be explained in terms of Darwinian evolution. For example, the hydroid and medusoid forms of the same individual coexist in many forms of jellyfish, corals, and hydras.

Williamson's prolonged study of marine metamorphoses has led him to disagree with Darwin's assumption that adults, embryos, and larvae have evolved in a progressive way along a single lineage. On the contrary, he proposes that all embryos and larvae originated by hybridization of two or more distinct species from quite different evolutionary groups. He suggests that this occurred either by the transfer of genes that coded for the larval stage or by the union of whole genomes. By hybridization he implies sexual union between the two radically different life forms, followed by amalgamation or swapping of whole genomic sequences. Further crosses between animals with and without larvae, at all levels of taxonomic relationships, may have led more and more groups to acquire a larval stage; some groups have three or even four metamorphoses in their development.

If Williamson is right, two or more complex animal genomes would have to coexist, in some modified form, in the same animal, with phases in which one inheritance dominates, followed by a switch to dominance by another.

The Genome's creative capacity has remarkable powers of self-government and control, as we know from the way all our human tissues and organs arise from a single fertilized cell. Williamson's preliminary sexual hybridization experiments in his laboratory confirm that at the very least hybridization between different life forms is possible, producing larval forms never seen before. He has predicted that many more examples of hybridization will be found, and he has urged geneticists to investigate these possibilities through nucleotide sequencing.

In ways that extend much deeper and wider than Gould and Eldredge imagined, saltations in the fossil record would appear to reflect actual changes in species over time, arising from multifactorial forces that often include symbiotic mechanisms. Such an approach does not mean that symbiosis is the sole mechanism of evolutionary change, any more than Darwinian gradualism is.

A single paradigm does not rule evolution or any other branch of science. Not only do multiple paradigms apply, they interact from event to event and from moment to moment.

· 20 ·

SEX AND CANNIBALISM

> His fifth film, which is closer to his real interest, shows mitotic cell division and the formation of male and female sex cells in this species. The scenes of fertilization are gorgeously sharp and clear. The male cell enters the posterior end of the female and fuses with her, his nucleus floats about, eventually contacts hers and they are one.
>
> — LYNN MARGULIS,
> "L. R. Cleveland: Scientist Misunderstood"

ON SEPTEMBER 20, 1519, the Portuguese explorer Ferdinand Magellan set sail from the port of Sanlúcar de Barrameda in his flagship, *Trinidad,* accompanied by four consorts, *San Antonio, Concepción, Victoria,* and *Santiago.* On board the vessels were 270 men destined to make the first successful circumnavigation of the globe. Of the five ships that embarked with pennants flying, only one, the *Victoria,* returned home, in 1522; Magellan was not on board, having been killed in a skirmish by natives in the Philippines. But brought back by the few surviving crew of the *Victoria* were two specimens of birds, which were duly presented to the king of Spain on behalf of the ruler of Batjan, one of the Molucca Islands. So gorgeous was the birds' plumage that the Spanish believed they must be visitors from paradise; today we still know them as birds of paradise.

The geographic isolation of New Guinea during the early Paleocene (about 50 million years ago) had allowed a family of unprepossessing crowlike birds to diverge into forty of the most colorful species on Earth. The male birds of paradise are the most decorative, but some display their magnificence in more subtle ways. One of these, known as Lawes's parotia, inhabits the forest of New Guinea. No larger than a crow and equally black, apart from a green breast shield and six attenuated feathers with bulbous finials sprouting from his forehead, the male reserves his display for courtship.

To begin with, he clears a stage on the forest floor under a convenient bower of trees, stripping all vegetation from an area about 15 feet in diameter. Then, having expanded his flank plumes into a top-shaped crinoline that sparkles with a beautiful turquoise sheen, he stretches his neck up to prominently display his green breast shield. Thus fluffed up in all of his finery, he performs an elegant and mesmeric dance, hopping from side to side. For the benefit of the females that have gathered to watch from an overhead branch, he embroiders the dance, pulsating his neck shield and rotating his head from side to side, so that the six long head feathers vibrate until the whole is lost in a dervishlike blur. In the blink of an eye, he darts up into the branch for a conjugal visit with one of his spellbound audience, a movement so slick and fast it is done before the female has a chance to react. Immediately he is back on the ground, once more performing his dance, puffing and prancing for visit after visit, as long as the spectacle holds his audience.

Sexual displays and courtship rituals are the most charming and amazing physical interactions in all of nature, and none more so than that between male and female of our own species. Beauty has never been entirely in the eye of the beholder; in significant part it is in the design of nature. But how, one wonders, could the evolutionary process that began with the humble bacterium or the first amoeboid cell have given rise to a spectacle such as this?

Of course, we know what sex is really about, beyond the choreographed displays of fashion, dancing, and music, beyond even the physical and spiritual pleasures and pains of the union; its primary purpose is reproduction.

The biology of sexual reproduction is the same in humans as it is in most other animals, plants, and the great majority of fungi, and even in many creatures that comprise one or very few cells. We are naturally interested in how sexual reproduction first came about. Yet the sexual union of male and female is not a prerequisite for reproduction. Life has much simpler means of achieving this.

Bacteria reproduce by binary fission, on average about once every twenty or thirty minutes. The green hydra simply buds off a new daughter from its main trunk in an asexual form of reproduction called parthenogenesis. Many plants can reproduce in this way, as can insects such as green flies and also a few species of reptiles, such as the lacerta of the Caucasus and the whiptails of Mexico and the American Southwest. The biological science of artificial cloning, which might appear to be parthenogenetic, is actually closer to symbiogenesis, since the nucleus of the new individual is transplanted into the cytoplasm of another cell, fusing the nuclear and cytoplasmic genomes. When life is produced from such a conception, it is wise to remember that both these genomes are as old as the cells they come from. And when the nucleus is transplanted into the cytoplasm of a different individual, this might give rise to incompatibility between the nuclear and mitochondrial genomes.

In fact, the commonest way in which nucleated cells reproduce is by the direct replication process known as mitosis. This is how the nucleus of the amoeba makes identical copies of itself. Our human cells, like those of all animals and plants, reproduce by mitosis, for example during the development of the embryo; it is also the way we replace cells that are damaged or naturally die off as part of the normal wear and tear in our tissues and organs.

Mitosis is a strange and beautiful process to witness through the lens of a light microscope. First the chromosomes make identical copies of themselves: the human genome, for example, increases its complement from forty-six to ninety-two. A spindle-shaped structure then appears in the cell, with a dotlike centriole at each pole and stringlike tubules radiating centrally from either centriole, meeting at an equator, where the chromosomes line up. Imagine a miniature Earth on which the centrioles are the north and south poles, which have pulled apart to form the points at opposite ends of the spindle.

The linking microtubules are forty-six lines of longitude. At each point where a line of longitude crosses the equator, the twinned chromosomes (now called chromatids) are attached by special structures called centromeres. Then all of these centromeres split in half, pulling their respective chromatids to the opposite poles. The result is two matching collections of forty-six chromosomes now gathered at the two opposing centrioles. The tubules and centrioles all melt away, and the nuclear membrane, which had disappeared, now reappears and pinches off at the former equator, forming two daughter nuclei, each destined for one of the two new cells, which are identical to each other and to the parent cell.

Mitosis is clearly an efficient process of reproduction. So why did life bother to evolve sexual reproduction — the evolutionary equivalent of Adam and Eve — with the complications and pitfalls of courtship and physical union? The very ubiquity of sex suggests that there must be a very powerful selective advantage, or advantages, for its evolution. Yet the biological role of sex and mating, with all of its accompanying displays and intrigues, is proving surprisingly difficult for scientists to explain.

One reason for thinking that sexual reproduction must have advantages is that it demanded a great deal of evolutionary adaptation. Consider the complex anatomy and physiology of sex in animals, for whom the act of consummation involves some danger. Rutting deer, like sexually competitive males of many other species, frequently injure each other in fighting over access to a mate. The whole purpose and being of the male great peacock moth, the largest in Europe, is nothing other than sexual reproduction. In the few days of its existence it does not eat, spending all of its time and energy in the search for a female, after which it dies. The female praying mantis sometimes devours the male after copulation, and the male redback spider, native to Arizona, which weighs no more than 1 to 2 percent of the average female's weight, will actually offer himself to the female's giant jaws, because in the time it takes her to eat him the proportion of eggs fertilized by his sperm increases. And pregnancy in placental mammals carries numerous risks to the female. For example, if the fertilized egg implants in the fallopian tube instead of the uterus, the mother's life is threatened. Without the advances of modern medicine, childbirth is hazardous, with a relatively high incidence of morbidity and mortality. How curious that the complex, dangerous, and energy-consuming processes of sexual reproduction not only evolved but came to dominate the reproductive strategies of both the plant and animal kingdoms.

For the moment let us put aside the complexities and ambiguities and look at what is going on from a Genomic perspective.

Current estimates place the number of species of plants and animals on Earth at more than 30 million. There is something aesthetically satisfying in the fact that, except for those that reproduce parthenogenetically, all of these species return to the ancestral state of a single-celled organism during sexual reproduction. This single cell arises from the fusion of sets of single chromosomes derived from the two parents. To enable this to happen, the parental germ cells must first undergo a reduction division of the chromosomes called meiosis. During meiosis the germ cells, or gametes, first abandon one set of their paired chromosomes — for example, in humans, the sperm and the egg contain only twenty-three of the forty-six chromosomes that make up the full genome — so that when the two gametes consummate their union, the fertilized cell is restored to its full complement. The stage at which the gamete has only a single set of chromosomes is called "haploid," and the normal or doubled stage is called "diploid."

Meiosis is a complicated performance that must have come about through an equally complex evolution. It differs from mitosis in a number of important ways. Where mitosis is an everyday happening in the organs and tissues of an animal or plant, meiosis takes place only in the sex or germ cells — in human terms, within the ovary and testicle. It begins with a duplication of the chromosomes similar to that seen in mitosis, but two cell divisions rather than one follow this. Prior to the first division, the four matching sets of chromosomes wrap tightly around each other, a prerequisite to their swapping corresponding segments. To understand what is happening, we need to recall that each parental cell has genes (called alleles) from two sources: the father's and the mother's germ cells. During the swapping, the alleles of corresponding segments are shaken up and put back together in a random assortment, much like the balls in a lottery draw. This results in considerable mixing of the parental genotypes and so becomes an important mechanism for genetic variation. Found only in meiotic reproduction, this mixing is known as sexual recombination, and it is the reason that, with the exception of identical twins, all of the children of the same parents differ slightly from each other.

In meiotic reproduction, as with mitosis, the spindle forms in the cell and the newly juggled chromosomes line up on the equator. Two versions of each chromosome line up side by side. But unlike what happens in mitosis, the centromeres do not split during the first reduction division. Instead, one of the two versions of the whole chromosome is drawn to each pole, and then the cell divides. The resulting cells have a full complement of chromosomes — in our case, forty-six. A second reduction division of the two daughter cells produces four new offspring, each containing a single (haploid) set of chromosomes. In human terms, each of these constitutes a sperm or an ovum, which, when united, restore the diploid set of forty-six chromosomes to the fertilized egg, or zygote.

Sexual reproduction is obviously far more complex, even at the genomic level, than asexual mitosis. Thanks to recombination, each individual zygote has a different blend of parental genes. Whether it will be male or female, for example, depends on whether it has an X or Y chromosome. Soon the zygote has become a ball of adherent cells called the blastula, the first recognizable stage of development in members of the animal kingdom. In plants the equivalent defining stage is known as the embryo. Soon the individual cells begin to change appearance and behavior, to become the forerunners of different tissues, such as muscle and bone, liver and brain. Yet — and this was an astonishing revelation to the biologists who first discovered it — every cell in every tissue still carries the identical genomic template that was formed in the zygote.

The fusion of the male and female gametes bears a striking resemblance to endosymbiosis. Of course in endosymbiosis the genomes of two different life forms merge rather than the genomes of two genders of the same species. Nevertheless, the similarities are so striking that, in the 1940s, Lemuel Roscoe Cleveland, a professor of biology at Harvard University, became interested in the possibility that meiotic sex might have first evolved as cannibalism between protists.

Cleveland spent many years studying the chromosomes of forty genera and more than five hundred species of the protists that live in the guts of termites and roaches. In these studies he came across many unusual examples of genomic change, from haploidy to diploidy and back again. He also noticed that in times of starvation, these tiny creatures would eat others of their own kind. This cannibalism was sometimes incomplete, so the cannibal ended up with several nuclei in its now bloated cell. The multiple nuclei sometimes fused into one, resulting in many copies of similar chromosomes in the same cell, a situation geneticists call "polyploidy." Polyploidy also occurred with surprising frequency from mistakes in mitosis, causing geneticists to worry that polyploidy could lead to genomic instability.

Cleveland wondered if the first meioses arose as a mechanism for reducing polyploidy. For mitosis to take place, the centromeres must divide and so allow the retracting spindles to pull their respective chromatids toward the centrioles. No such splitting of the centromeres takes place during the first reduction division of meiosis, which seemed relevant to Cleveland. He also noticed what appeared to be accidental mistiming in the division of the centromeres. Adding that phenomenon and the cannibalistic polyploidy, he theorized that a series of mishaps during mitosis might have given rise to the first clumsy version of meiosis.

Sadly, when Cleveland described his findings to fellow scientists in 1947, they were not very receptive. An engaging if rather irascible man, Cleveland entertained his audience with films of his observations, narrating in a rural Mississippi accent; nobody took him seriously. Later on, however, Lynn Margulis did take him seriously, and it is to her credit that we know about this interesting scientist and his experiments. In her words, "I worked with a big blue ciliate, called *Stentor coeruleus*, for maybe ten years and I used to see the stentors eating each other. When things got bad, many encysted. But those that couldn't encyst turned to cannibalism as a way out." Today Margulis believes that sex may have begun as an endosymbiotic union between different species of protists. Cleveland's hypothesis also fits neatly with Margulis's observation that the first protists contained only one set of chromosomes. With the genomic union of two such haploid protists, the holobiotic cell became diploid.

By 700 million years ago, the first soft-bodied animals were engaging in sexual reproduction, leading scientists to suggest that meiosis evolved about a billion years ago. However it happened, whether through cannibalism or some other mechanism, there was a terrible price to be paid for the abandonment of mitosis.

Bacteria are not programmed to die. They will keep on dividing forever unless some outside agency, like a chemical poison or ultraviolet light, destroys their metabolic processes. But meiosis programs all creatures that employ sexual reproduction to die. "Death," as Margulis graphically expresses it, "is a sexually transmitted disease."

In a subtle sense, however, she is not completely correct because each organism can pass on half its genome to potential offspring. Each new individual inherits genes from two parents, who in turn inherited their genes from four grandparents, and so on. Every individual has not only received genes from many ancestors, but will, assuming reproduction, pass on those genes to many future descendants. In this way the species as a whole is an evolving unit, with a shared gene pool. It is possible, though not wholly comforting, to look at each individual human being as a "superorganism" that proliferated through cloning from a single cell, the inheritor of its genes. The genes live on as long as future generations exist to inherit them.

While Margulis doubts that meiotic sex adds anything useful, other than stability, to the evolving genome, Darwinians such as John Maynard Smith are inclined to disagree.

Various theories have been put forward to explain the advantages of sexual reproduction; for example, it creates a greater diversity of individual genomes within a species. This diversity might improve the chances of individual survival in a rapidly changing, or threatening, environment — as when a highly lethal epidemic infection threatens a population. Our human genome contains many viruses and fragments of viruses, which suggests that genetic diversity did play a part in human survival through previous virulent pandemics. Thus, as Edward O. Wilson succinctly says, "Diversity is the way a parent hedges its bets against an unpredictably changing environment."

Some evolutionists believe that sexual reproduction may be maintained by natural selection working not at the level of the individual but at the level of a group of organisms. Group selection, they claim, allows for a more rapid evolution of subgroups within a population to meet new environmental challenges. At a genomic level, as Smith points out, sexual reproduction might also reduce the effect of harmful mutations: a genome created by meiotic sex has two copies of every gene, so a harmful gene, if not dominant, would be rendered harmless by the matching normal gene. Other arguments for the usefulness of sexual reproduction include genomic efficiency in repairing chromosome damage and the control of selfish behavior by individual genes.

Smith and Szathmáry have put forward their own Darwinian scenario for the evolution of meiotic sex, which envisages three stages, beginning with a haploid-diploid-switching life cycle, continuing with an early form of sexual meiosis, and further to the full modern sexual cycle, which allows for crossover of genes before the formation of new gametes. Such a hypothesis could easily accommodate Cleveland's protist fusion in its first two stages.

Whatever the reason for the evolution of sexual reproduction, it heralded the most colorful revolution in the history of the biological world. Sex is the reason that the butterfly orchid displays its petals and the lily of the valley exudes its rich fragrance in late spring; it is the inspiration for the male peacock to spread his glorious feathers and for mating grebes to literally walk on water. Sex is also the reason why young girls faint at rock concerts, as it was the inspiration of so many poets, including Donne and Shakespeare, painters such as Fragonard and composers such as Verdi, evoking some of the most enchanting compositions in all of the "divine" arts. Indeed, the delights of courtship and romance must surely have played an important part in the evolution of the aesthetic sensibility that helped make us human.

We should not forget, in all of this wonder, that the simplest form of reproduction, asexual binary fission, remains an integral part of our human identity: it is the way our mitochondria still reproduce, more than a billion years after their first integration into the eukaryotic cell.

Evolutionist Richard Law believes that mutualistic symbiosis tends to protect the partnership by resistance to further evolutionary change. If he is right, this resistance is more readily maintained through asexual rather than sexual reproduction. This would explain why many long-term mutualistic symbioses, such as the incorporation of mitochondria, involve asexually reproducing organisms.

How astonishing it is to reflect on the history of these minuscule partners — where they came from, how long ago, and how they link us, and all of visible nature, to that common eukaryotic ancestor.

The vast time that has elapsed since their first endosymbiotic incorporation has given rise to an extraordinary diversity of mitochondria in different eukaryotic life forms. The human mitochondrial genome is the smallest of all, retaining only thirty-seven of the original three thousand or so ancestral bacterial genes, a quarter the size of the mitochondrial genome of yeasts. This functional compaction has no parallel in the biological world except in the viral genome.

The endosymbiotic evolution of mitochondria has other important applications. All mitochondrial DNA in a human being is inherited solely from the mother, since the cytoplasm of the fertilized egg comes from the ovum. The sperm sheds its small complement of mitochondria (which provide energy for the whiplike tail) when it penetrates the ovum. Mitochondrial DNA mutates at a predictable rate. By comparing the sequences of mitochondrial DNA, one can estimate the closeness, in evolutionary terms, of different species of animals and plants. And within a single species, it is possible to estimate the closeness between groups separated by time and by geography. By comparing mitochondrial DNA from different human groups, scientists have arrived at some remarkable conclusions.

Svante Pääbo is a Swedish geneticist working at the Max Planck Institute in Germany. In the 1990s he extracted small amounts of mitochondrial DNA from the fossilized remains of a 30,000-year-old Neandertal found in a cave in the Neander Valley in Germany. Pääbo then compared some of the mitochondrial DNA sequences of the Neandertals with those in modern humans; he found far too many differences to view Neandertal as a direct ancestor of our species. This mitochondrial evidence gave weight to what many paleontologists already believed: that the Neandertals were an evolutionary dead end, an extinct species.

Similar studies of human mitochondrial genomes in many different ethnic cultures and skin colors have, like the Human Genome Project itself, confirmed that we are all quite closely related to each other. Eighty percent of mitochondrial DNA variations occur within populations rather than between continents. The greatest variation within a single population was found in the Bushmen of southern Africa, who must therefore have been around as a distinct group for longer than anybody else. This has led anthropologists to speculate that the Bushmen represent the closest people on Earth to our true ancestors.

Perhaps most surprising of all, studies of mitochondrial DNA also suggest that every human being on Earth is descended from a common maternal ancestor, the so-called African Eve, who lived roughly 140,000 years ago. Human evolution did not really begin with this postulated Eve: many hominid ancestors evolved earlier in Africa, including *Homo erectus,* which left a global fossil record that tracks its diaspora to at least 1.5 million years ago. In Europe this hominid, after some intermediate stages, is thought to have evolved into the Neandertals. Meanwhile, in Africa, a parallel but distinct line of descent led eventually to the suggested Eve.

Of course, our evolutionary origins go back a great deal further than early hominids, yet it is surprising to learn that after more than a billion years of symbiotic incorporation, with the original bacterium losing much of its genetic identity, our human mitochondria still retain some independence of behavior from the host cell. This has important consequences.

One is that, like their bacterial ancestors, our mitochondria are sensitive to certain antibiotics, a factor that must be considered in the development of new antibiotic drugs. In another example, Christoph Schmitz and his colleagues at the University of Aachen have shown that exposure of the human fetus to modest doses of x-rays may lead to subsequent mental illness. The researchers noted that children who were born within nine months of the Chernobyl disaster had an increased risk of abnormalities in the part of the brain called the hippocampus, and they were more likely to have subsequent behavioral problems. To test this, they exposed pregnant mice to the equivalent of ten medical x-rays. Six months after birth, the offspring showed significant loss of cells in the hippocampus area. The researchers attributed this loss to damage affecting the mitochondria, which could not provide enough energy to repair the DNA damage brought on by the irradiation.

As they have from their earliest endosymbiotic amalgamation into the eukaryotic cell, our mitochondria serve as the cell's power supply. A delicate arrangement of five enzyme complexes in the mitochondrial inner membrane supplies the energy according to the requirements of the different tissues and organs. The numbers of mitochondria per cell vary enormously, depending on the tissue or organ's needs. For example, muscle cells need more mitochondria than most, and the mitochondria in heart muscle take up a third of the entire volume of the cells.

As a female fetus develops, about 80 percent of the eggs in the ovaries are culled, and some scientists believe that the culling may, in part, be a programmed elimination of eggs in which the nuclear and mitochondrial genomes are not perfectly compatible. Other researchers have gone so far as to suggest that reducing the mutational load of mitochondrial genes may be part of the reason for the evolution of sexual reproduction in the first place. In keeping with this suggestion is evidence that mitochondria in mammalian sperm are actively destroyed after a sperm cell fuses with the ovum.

One of the most important developments in recent research has been the recognition that mitochondria play a central role in "programmed cell death," or apoptosis. In the normal body processes, aging cells are killed off so they can be replaced by younger, more vigorous copies. Mutations in mitochondrial genes may also play a critical part in the aging process, as well as in human illnesses, such as certain neuromuscular diseases, cerebral and optic nerve degeneration, Parkinson's disease, Alzheimer's dementia, and some cancers. Meanwhile, specific mitochondrial gene variants have been found to be associated with longevity.

This continuing legacy of our symbiotic origins may even play a part in a new era of human exploration: the colonization of our solar system. In the weightless conditions of space, even small cuts in the skin do not heal properly, a problem that has complicated the lives of astronauts. In the words of Harry Whelan, a neurologist at the Medical College of Wisconsin in Milwaukee, "The reason for this is not well understood, but a cell's mitochondria — its energy sources — don't function efficiently in zero gravity, and this leads to a variety of health risks." Fortunately, the situation returns to normal when the astronauts return to Earth.

· 21 ·

PLAYING HARDBALL WITH EVOLUTION

We all play hideous games with each other. We step inside each other's chalk circles.

— ANTHONY HOPKINS

ON WEDNESDAY, SEPTEMBER 21, 1994, I stepped out of a yellow cab opposite Rockefeller University in New York. I had arrived a little early for my interview with its world-renowned president, Joshua Lederberg. It was unseasonably hot, so I walked along Sixty-eighth Street, toward the cool of the East River. I welcomed the opportunity to gather my thoughts, for I suspected that the forthcoming interview would change, quite radically, the way I viewed the evolutionary process.

Lederberg's contribution to science did not stop when he and Edward Tatum published their paper on microbial genetics in *Nature* in 1946. Six years later, with Norton D. Zinder, he reported that genes could be exchanged between bacteria with the help of viruses, a process they termed "transduction." So important was this finding that, in the words of the *Encyclopaedia Britannica,* "Lederberg's discoveries made bacteria as important a tool of genetic research as the fruit fly, *Drosophila,* and the bread mold, *Neurospora.*"

By the time of my visit, viruses were proving an invaluable tool for scientists who wanted to alter genomes. Lederberg had confirmed that nature got there first. When a virus enters a genome, it may bring in foreign genes. First observed in viral infections of bacteria, this discovery was beginning to stimulate a much broader interest in the evolutionary role of viruses. And so, for an entire afternoon, I was privileged to enjoy the clarity of Lederberg's thinking and articulacy of expression as we talked about the pluripotent nature of viruses.

He explained that in 1952 he had introduced the word "plasmid" to denote any mobile genetic package as a unifying concept in cell genetics and hereditary symbiosis. At that time the role of cellular organelles and some other particles outside the nucleus was controversial; opinions differed on whether they were symbiotic or merely parasitic. There was heated debate as to whether or not an infecting virus could be considered genetically important. As Lederberg explained, "The notion of infection was antithetical to the view that a virus could be a hereditary factor. I said there was no operational distinction between them. The same particle can be an infective agent or a hereditary unit."

Viruses, in other words, could affect the evolution of the hosts they infected. This was a message I conveyed in *Virus X*, in which I looked at ways that even plague viruses might play a formative role in evolution.

When one accepts that viruses are alive, it becomes clear that any virus infection of a reproductive cell, whether of a single-celled life form such as an amoeba or a germ cell of a plant, animal, or fungus, is inevitably endosymbiotic. There could be no more definitive example of living together, for a virus becomes metabolically alive only when it enters the genome of its host.

It may be that not all scientists are familiar with Lederberg's thinking; even today, some react suspiciously to this more enlightened view of viruses, convinced that they are no more than parasites. This attitude mirrors the resistance that met symbiologists' claim a generation ago for the evolutionary role of bacteria. Some biologists still fail to grasp that parasitism is a form of symbiosis, one extreme over a wide range of living interactions. We have seen many examples of mutualistic symbiosis that began as parasitism, with no easy junction to define where parasitism ends and mutualism begins. As René Dubos and others have repeatedly explained, the symbiotic interaction is dynamic, as capable of change as Darwinian selection pressures; a physical characteristic that was an advantage yesterday may become a disadvantage tomorrow, in evolutionary time.

Viruses have many properties that make them of exceptional evolutionary interest. They are the only life forms to inhabit the genome. Not only are they well adapted to infect large numbers of individuals of a host species, they are also capable of crossing the species barrier. The great pandemic of human influenza that swept the world after World War I resulted from human infection with a hybrid virus that had arisen from the endosymbiotic union of two different strains that most likely originated in ducks and pigs. The capacity to blend genomes of different strains (species), combined with their facility to jump from one host species to another, places viruses in a unique position to influence and even to radically change the evolution of their hosts. Consider also that every species of life on Earth has viruses that infect it. The more one understands about viruses, the more impressed one is by their evolutionary potential.

Equally impressive is the range of viral behaviors once they enter the genomes of their hosts. Symbiosis, as I have repeatedly emphasized, does not imply some cozy partnership. Even as mutualism, it involves hard bargaining that can, and often does, begin with brutally selfish exploitation.

The phylum Craniata includes a variety of classes, from cartilaginous fish, such as rays and sharks, to the bony fish, amphibians, reptiles, birds, and mammals. All mammals have certain characteristics in common, for example a large brain for their body size, warming hair or fur that covers the skin, and the provision of milk to nourish offspring after they are born. The mammals are divided into two subclasses: those that lay eggs, such as the spiny anteater and duck-billed platypus, and those that give birth to offspring mature enough to suckle milk. The latter subclass includes the marsupials and, arguably the most successful of all, the placental, or eutherian, mammals, to which humans belong. One of the greatest leaps in mammalian evolution, and the development that gave the eutherian mammals their survival edge, was the ability to nurture the developing fetus within the maternal body, in the chamber we call the uterus or womb.

No other life form on Earth has evolved this complex gestational arrangement, which requires the solving of major difficulties. Uterine gestation carries obvious risks to survival. The human fetus inherits 50 percent of its antigens from the father, many of which are "foreign" to the maternal immune system and would normally prime the mother's system to attack the fetus. This form of gestation required the concomitant evolution of the placenta, a remarkable organ that might be compared to a buffer zone between two interacting circulations of mother and fetus; it extracts nourishment for the fetus from the mother's blood while protecting the fetus from the potentially lethal attack by the maternal immune system.

Indeed, the evolution of the placenta is critical to understanding how the eutherian mammals came to exist. In the early 1970s a group of scientists was studying the placenta of baboons to find out whether the rubella virus could cross from the maternal to the fetal circulation. Rubella, otherwise known as German measles, is notorious for causing fetal abnormalities, and the researchers were studying how such catastrophes might be averted. Scanning through landscapes of baboon placental tissue under the electron microscope, the researchers, including J. Robin Harris, from the Institute of Zoology at the University of Mainz in Germany, readily identified viruses on their grids; but they were startled to observe large numbers of tiny spherical viruses that were not rubella at the junction of the placental surface and the lining of the womb. Equally surprising was that the researchers found these viruses in what appeared to be perfectly healthy baboons. Similar viruses were soon detected in the placentas of cats, mice, guinea pigs — and humans.

The viruses these researchers found, known as retroviruses, are among the most amazing life forms on Earth. They are also a little scary from an evolutionary perspective. With a genome based on a template of RNA rather than DNA, retroviruses are so called because once inside the host cell they use their own enzyme, known as reverse transcriptase, to convert the viral RNA into its matching DNA before the viral genes are incorporated into the nucleus of the host. Retroviruses exist in two different forms. *Exogenous* retroviruses infect their hosts from the surrounding environment. The best-known example, HIV-1, the cause of AIDS, is acquired through sexual intercourse, sometimes through pregnancy, or through injection with contaminated needles and syringes. *Endogenous* retroviruses, like those the scientists found in the baboon placentas, insert their transcripted DNA into the reproductive cells of the infected host, where it is incorporated into the genome and passed down, along with all of the other genetic coding, from generation to generation. Some of the retroviruses in our human genome have been there for millions of years.

Today a growing number of microbiologists, including Erik Larsson, at the University of Uppsala, believe many of these endogenous retroviruses play an essentially symbiotic role. In 1988 Larsson suggested that the retroviruses in placentas may actually help protect the fetus. Certain viruses have the capacity to fuse mammalian cells into confluent sheets of cytoplasm with many nuclei and no cell membranes between them. Multinucleated "giant" cells are a well-known feature of AIDS, probably helping the HIV-1 virus to spread. Fusion of cells is also a feature in the mammalian placenta, with the formation of a microscopically thin and confluent tissue layer, called the syncytium, that is the final barrier between maternal and fetal circulations. All nutrients, immune cells, and antibodies from the mother to the fetus must pass through this syncytial layer. There is reliable evidence that endogenous retroviruses play an important role in the structure and function of the syncytium.

Luis P. Villarreal, a professor of molecular biology and biochemistry at the University of California at Irvine, is a globally acknowledged expert on viruses and their evolutionary implications. When Villarreal tested Larsson's hypothesis by inserting into the cells of a mouse embryo a virus known to suppress endogenous retrovirus genes, nothing adverse happened. But when he put the suppressive virus into cells that form the placenta, the placenta no longer implanted into the mouse uterus, derailing the process of uterine conception.

Neil Rote, a reproductive immunologist at Wright State University in Dayton, Ohio, has demonstrated an even wider role for endogenous retroviruses in placental physiology: they control certain aspects of cellular differentiation, inducing cells to change into syncytium. These findings suggest that in the evolution of the placental mammals more than 100 million years ago, the retroviruses would have prevented the mother's immune cells and antibodies from rejecting the fetus. And Joachim Denner, working at the Paul Ehrlich Institute in Langen, Germany, has found in the placenta a retrovirus that is a potent suppressor of this type of immune reaction.

J. Robin Harris poses an even larger question: "Could an ancient endogenous retrovirus . . . be responsible for the creation of the placenta?" Harris concludes that a definitive answer is not yet possible but may emerge from further analysis of the evolutionary lineages of various animal and human endogenous retroviruses.

This viral role has been gathering support from detailed studies in many laboratories. The most important evidence to date has come from the research of Sha Mi, John McCoy, and their colleagues at the Genetics Institute in Cambridge, Massachusetts, who identified one of the viral genes responsible for the formation of the syncytium. Normally coding for an envelope protein of an endogenous retrovirus, this gene had been incorporated into the human genome to code for a protein they called syncytin, which plays an essential role in the formation of the interface between the placenta and uterus.

The condition known as preeclampsia in pregnant women is characterized by fluid retention and high blood pressure. Uncontrolled, it can lead to eclampsia, which can threaten the life of both mother and unborn child. In 2001 Sha Mi and her colleagues found evidence that preeclampsia is accompanied by a dramatic reduction in the expression of the syncytin gene and the apparent dislocation of the protein from its normal position in the syncytial tissues. The following year I. Knerr, E. Beinder, and W. Rascher at the University of Erlangen-Nuremberg in Germany confirmed that preeclampsia was associated with low syncytin levels in the placenta. Perhaps in time further research in this area will help to prevent or even treat this common and important illness.

The virus that produces syncytin, known as HERV-W, is located on chromosome 7 in the human genome. When Sha Mi and her colleagues first searched for this gene in nine other species, including representatives of several other mammalian orders, they found similar genetic sequences only in the rhesus monkey. This appeared to cast doubt on the idea that syncytin played a formative role in the early evolution of placental mammals. But subsequent testing by McCoy turned up similar genes in mice. In a personal communication, McCoy explained, "We probably missed them in the test system we used because the sequences have diverged quite a bit during evolution." There would have been a great deal of time for evolutionary divergence to take place, since the origins of rodents go back about 60 million years — much closer to the true beginnings of placental mammals.

Like the mitochondria in the cytoplasm, this virus has shed many of its original genes during its long endosymbiotic relationship within the mammalian genome, so it is classed as a "defective virus." This means that it can no longer multiply without the assistance of other viruses, but its genes continue to cooperate with the rest of the genome, playing what appears to be a fundamental part in human reproduction. These findings are part of an exciting new wave of research into the role of viruses in the evolution of their hosts.

As a graduate student at the University of California in San Diego in 1976, Luis Villarreal studied how some viruses, rather than infecting and killing their host, use a strategy of long-term persistence within the genome. After spending many years investigating this, he realized the limitations of Darwinian thinking about the relationship between viruses and their hosts. How, for example, did one rate the "fitness" of an organism that reproduces little, if at all, over a million years of inhabiting the genome of its host? In neo-Darwinian thinking, these viruses have no fitness, yet their strategy of persistence makes them virtually immortal.

The conventional theory assumes that viruses acquire genes by stealing them from their hosts, but Villarreal believes that viruses are enormously creative in themselves. They evolve up to a million times faster than their hosts, tolerating high rates of replication error while stitching together bits and pieces of hereditary material to make new genes. One such gene invented by viruses codes for what is known as the tumor antigen, or T antigen. Although in humans it may cause tumors, in the viral life cycle this antigen interacts in complex ways with fifteen different host nuclear systems, including the DNA replication apparatus. There is no equivalent gene in any other life form.

In February 2001, Villarreal made an iconoclastic presentation to the annual meeting of the American Association for the Advancement of Science. He began:

> Persisting viruses can have major evolutionary consequences for the infected host, causing rapid change in its physical makeup . . . Such events seem to mark major breaks and new orders, such as the development of the eukaryotic nucleus or of the adaptive immune system . . . Yet these punctuated acquisitions would not appear likely or even feasible based on conventional neo-Darwinian models for evolution that emphasize point mutations and sexual recombination of host genes to create novel phenotypes.

An earlier generation of neo-Darwinians assumed that the numerous viral genes in the human genome were "junk." But a more considered reappraisal in recent years has made geneticists more cautious in their assumptions. Retroviral genes have recently been linked to genetic susceptibility to important groups of illnesses, including cancers, leukemias, and autoimmune diseases such as lupus, rheumatoid arthritis and, perhaps, multiple sclerosis. In 1996, in a review of the roles played by viruses in humans, Roswitha and Johannes Löwer and Reinhard Kurth of the Paul Ehrlich Institute listed their many different potentials for genomic change. Persisting endosymbiotic retroviruses, which they assumed must derive from germ cell infections in the ancient past, are being discovered in a growing range of animal phyla, from sea urchins to humans, making it likely that viruses have been involved in the evolution of many, and possibly all, species. While some endogenous retroviruses were incorporated into human germ-cell lines after our ancestral separation from other primates, many are so ancient that we share them with Old World monkeys. And once viruses enter a genome, their capacity for evolutionary novelty remains: even millions of years later, they can interact with newly arrived viruses. In this and many other ways, viruses increase the "plasticity" of our genome, enhancing its potential for change at any stage.

Some viral genes have become incorporated into our coding for useful metabolic and physical functions, even enhancing our immunity to other viruses, including other retroviruses. If Villarreal's theory is right, viral infection of the ancestor of all the eukaryotes has made a fundamental contribution to the most important evolutionary transition of all: that of the eukaryotic superkingdom.

Vital to the chemistry of DNA is a group of enzymes known as DNA polymerases, which make possible the copying of DNA, a role of such critical importance that Villarreal and his colleague Victor R. DeFilippis believe that the genes that code for these polymerases have been conserved throughout evolutionary history. Yet when the scientists compared the DNA polymerases of bacteria and eukaryotes, they were surprised to find that a key family of enzymes, known as the "B-polymerases," differed greatly between these two kingdoms. According to SET, the eukaryotic cell is descended from a series of bacterial mergers. But symbiologist Margulis as well as the Darwinians Smith and Szathmáry have been baffled by the evolution of the nucleus. Villarreal and DeFilippis now confirmed that the B-polymerases in eukaryotes, including humans, are strikingly similar to those found in certain viruses, such as the Epstein-Barr virus, the cytomegalovirus that causes glandular fever, and a kind of virus that infects bacteria, known as the T4 phage virus. This raises a question of great evolutionary importance: might a DNA virus (one whose genome is composed of DNA) that "infected" the early forerunner of the eukaryotes have donated the original gene or genes necessary for DNA replication to all subsequent eukaryotic life? In his lecture, Villarreal explained why he now thought this likely.

Villarreal's viewpoint cannot be dismissed. A formidable expert in the molecular biology of viruses, he has written the chapter on viral evolution in the international bible of virology, *Field's Virology*, and an immense amount of original research and experience supports his idea. Thus for Villarreal and DeFilippis, a viral origin is not such a surprising explanation for the B-polymerases so vital to all subsequent eukaryotic evolution.

If these researchers are right about another related theory, viruses have contributed much more to cellular evolution than a single family of genes, however important. Scientists estimate, from analysis of DNA sequences, that the universal common ancestor of all eukaryotic life started off with a surprisingly small nuclear genome, consisting of about 360 genes. This is the size of a large DNA virus, and biologists such as Philip Bell of Macquarie University in Sydney have suggested that the eukaryotic nucleus began as an infection of a bacterial cell by a DNA virus. This view, which Villarreal shares, is not far-fetched. In fact, it is as convincing a theory as any that has been proposed to date.

According to SET, an endosymbiotic union of bacteria gave rise to a prototypical cell that included the cytoplasm, mitochondria, chloroplasts, and, if Margulis is right, the motility apparatus, including cilia and centromeres. Incorporation of a DNA virus at some key stage would have kick-started the nucleus. The many commonalities between DNA viruses and the nucleus make the scenario plausible. Certain DNA viruses have double-stranded DNA that is bound in a double membrane, just like the nucleus. While the genes of most archaebacteria and eubacteria are bound up in a circular molecule, these DNA viruses have linear genes like those found in eukaryotes. Viruses readily incorporate their DNA into other genomes, so the eukaryotic cell could readily have originated from the blending of viral and bacterial genomes, including SET. Viruses also tend to keep the replication of DNA separate from its translation to proteins, again a feature of the nucleus-cytoplasm dichotomy in the eukaryotic cell. Viral DNA is wrapped in a protein very similar to that of nuclear genes, and the viral RNA is capped in the same way as the RNA in our human cells. Even the fusion of membranes that takes place when the sperm fertilizes the egg resembles a viral mechanism for cellular penetration. If one also considers Villarreal's evidence for a viral origin of the eukaryotic DNA B-polymerase enzymes, a viral origin of the nucleus becomes even more plausible. Although some biologists might claim that viruses stole these mechanisms from the genomes they once infected, Villarreal has accumulated a weight of evidence that points to the contrary. Others claim that viruses could not have evolved until the cells they infect had already evolved, but this too is an erroneous assumption. In test-tube experiments probing the origins of the first self-replicators, selfish elements quickly emerge to parasitize the parental self-replicators. The same results are seen in computer simulations based on self-replicating algorithms. This suggests that protoviruses began with the very dawn of protolife.

In a forthcoming book, *Viruses and the Evolution of Life,* Villarreal and his colleague Esteban Domingo make a detailed and compelling case not only that viruses have played a role in evolution overall but more specifically that the eukaryotic nucleus is viral in origin.

These revelations highlight the enormous evolutionary potential of viruses. It is revealing, in this context, that scientists have estimated that 10 percent of the human genome is DNA that was incorporated from retroviruses. The human genome contains a thousand complete endogenous retroviruses and tens of thousands of viral fragments. Human chromosome 21 alone contains two thousand viral fragments. One wonders how so many viral genes found their way into the coding apparatus of a single species; whether each endogenous retrovirus represents some distant pandemic; and what proportion of the persisting viruses represent coevolutionary symbiotic partners. While the *final* relationship between virus and host sometimes may be a mutualistic bargain, we should not forget that the relationship is malleable and that the *first* contact between the interacting species can also have evolutionary significance; however, the evolutionary mechanism involved must have been the hardest ball game ever played out in the none-too-gentle amphitheater of life's history.

In the 1950s a myxomatosis epidemic was deliberately engineered to exterminate the enormous rabbit populations in Australia. The scientists who conducted this act of biological warfare took samples of a virus symbiotic with the Brazilian wood rabbit and injected it into feral rabbits in southeast Australia. What happened next was very similar to what Kwang Jeon observed in his amoebae infected by the plague bacterium. A pandemic struck a vulnerable species, but, as with the amoebae, not all of the rabbits succumbed. Two of every thousand rabbits had a greater resistance to the virus and survived. In time a new relationship emerged, with rabbit and virus living together in a permanent symbiotic partnership.

This sequence, of culling followed by coevolution, has obvious evolutionary implications. According to Smith, it is the best "evolutionarily stable strategy"; by not destroying the host, the parasite maximizes its reproduction and transmission to other individuals among the host species. But Darwinians ignore an additional implication. If a resistant rabbit carrying its myxomatosis virus partner came into contact with a virgin population of rabbits from its parental lineage, the virus would still destroy the parental lineage, thus improving the chances of survival of the partnership. For this vicious evolutionary ball game, I have coined the term "aggressive symbiosis."

Aggressive symbioses are not confined to virus-host relationships. Any biting or stinging insect, predatorial animal, or pathogenic microbe, whether bacterium, protist, fungus, or virus, may enter into a symbiotic partnership in which its host offers food, or its body as shelter, in return for the advantage afforded by its symbiont's aggressive behavior. Certain ants that live inside acacia trees protect their host by attacking herbivores that attempt to browse the trees. Fungi of the order Clavicipitales live inside the leaves of grasses and produce toxic alkaloids that poison herbivores attempting to eat the grasses. A fascinating and deadly example of aggressive symbiosis is the association between species of luminous bacteria and a nematode worm. The symbiosis is complex, with the juvenile worm nurturing the bacteria inside it until it locates a suitable caterpillar prey. The worm enters the prey through the anus, mouth, or other body opening and penetrates the caterpillar's vascular system. The worm voids the bacteria from its intestine, and the bacteria attack the caterpillar, often killing it within hours. Both worm and bacteria multiply inside the dying caterpillar, each playing a part in digesting its internal organs. The new generation of worms ingests the bacteria once more, ready to begin another infective cycle. Meanwhile, the bioluminescence of the bacterium, while still inside the empty husk of the caterpillar, helps to attract new prey coincident with the emergence of the new generations of worms.

Ichneumon wasps, whose larvae are parasitic on the caterpillars of other insects, provide an even more startling example. The larvae can survive in the host caterpillar only in the presence of a virus that replicates in the ovary of every female wasp; as the egg is injected into the body of a caterpillar, it is wrapped in a liquid shell teeming with viruses. The virus then multiplies in the prey, suppressing its cellular immunity and enabling the larva to eat it from the inside. This viral role in protecting the eggs of its symbiotic partner bears some resemblance to the viral assistance with implantation and survival of the human embryo within the mother's womb.

HIV-1 is another viral example. Originating in chimpanzees in the African rain forest, the virus coevolves with its primate partner without causing any illness, but for humans it is almost as lethal, if untreated, as the myxomatosis virus was for Australian rabbits. Consider what would have happened if the HIV-1 virus had spread by a more contagious route, for example by sneezing and coughing, into the global human population.

As I explained in *Virus X*, viral culling — indeed epidemic culling of any sort — selects for a specific genetic type within a species. With the majority of other genetic types killed off, the type selected by the microbe can proliferate and spread, thus becoming the common "genotype" of the host population. Every form of life on Earth has viruses that coevolve with it in this way. Terry Yates, professor of zoology at the University of New Mexico in Albuquerque and a foremost expert in mammalian taxonomy, believes that the ongoing evolutionary partnership between viruses and their hosts is so close that if he knew the genome of a virus that coevolved with the duck-billed platypus, he could place the platypus itself in its correct spot in the evolutionary tree.

Plague culling, most likely viral, offers one possible scenario for the postulated single common ancestor of our modern human species,

the "Out-of-Africa" or "Eve" scenario. Our human ancestors inhabited areas close to the great African rain forest and the surrounding savannas, the source of many of the most dangerous emerging infections in history, including Lassa fever, yellow fever, malaria, HIV, and Ebola. In the cases of malaria and yellow fever, the means of transmission are also abundantly present, in biting mosquitoes. Because the African rain forest and savannas are also home to many primate, rodent, and bat species, with their coevolutionary viruses, emerging viruses with the potential to cause devastating human infection would inevitably have been a threat throughout our ancestral African evolution. If one assumed a vulnerability corresponding to that of the Australian rabbit to myxomatosis, a hominid population of, say, 30,000 to start with might be reduced to a mere ninety surviving individuals in a single epidemic. Given the scattered and migratory nature of early hominids, multiple epidemics involving slow-spreading viruses, whether insect-borne or sexually transmitted, would have been just as selective of the species as one major catastrophe.

The reality may well have been more complex, with climate change or scarcity of food resources leading to the exposure of migrating or displaced populations to hitherto unknown epidemic agents. Of course these are no more than conjectures, but they would explain why, after a long evolution that must have resulted in much larger populations, every human being on Earth is closely related to a single, not-too-distant ancestor — the mitochondrial Eve.

If a current hypothesis proves correct, symbiotic retroviruses may have helped to make us "human" in an even more remarkable way. Caves in France and Spain, like rocks in South Africa and Australia, are decorated with paintings of animals, figures, and abstract forms that date from more than 30,000 years ago. The paintings, like some archaic fertility sculptures, combine a wonderful accuracy of observation with fine aesthetic appeal. David Horrobin, the medical adviser to the Schizophrenia Association of Great Britain, believes that the genetic factors responsible for mental illness are intrinsically linked to the intellectual evolution that is revealed in this art and that made its cultural evolution possible.

In 2001, Horrobin published *The Madness of Adam and Eve: How*

Schizophrenia Shaped Humanity, in which he explains how schizophrenia, in individuals and their families, may be associated with the most intelligent and imaginative members of our species. Schizophrenia has blighted the families of many geniuses, including Albert Einstein, whose son was affected, as was the daughter of James Joyce and the mother of Carl Jung. Psychiatrists have long been aware that other members of a family in which an individual is schizophrenic may exhibit lesser features of the same personality and behavioral disturbances, or "schizoid" tendencies. Horrobin details the long list of geniuses who had schizoid tendencies, including Robert Schumann, August Strindberg, Edgar Allan Poe, Franz Kafka, Ludwig Wittgenstein, and even Isaac Newton. People with such mental illnesses make unexpected, often inappropriate, connections between day-to-day events, raising the possibility that schizophrenia, with its curious link to human genius, may have imbued certain of its sufferers with the creative insight to lift our human cultural evolution to new intellectual levels. These, according to Horrobin's hypothesis, include the beautiful Paleolithic cave paintings in France and Spain.

Horrobin assumes that schizophrenic tendencies arose through the Darwinian mechanism of mutant genes under selection pressure. But other experts have suggested a much more shocking alternative. Just as Horrobin's book was being published, Robert Yonken of Johns Hopkins Children's Center in Baltimore and a group of Swedish researchers, including Hakan Karlsson, of the Caroline Institute, Stockholm, found evidence of a retrovirus in brain tissue and cerebrospinal fluid in ten out of thirty-five people who developed acute symptoms of schizophrenia. The same virus was found in only one of twenty people with chronic schizophrenia and in none of the people tested as control subjects. As with any new discovery, this finding needs to be confirmed by other studies. If it is confirmed, the implications for viral symbioses with humans will extend far beyond mental illness.

However strange it might seem, we are descended not only from bacteria; we are, in part, also descended from viruses.

Again and again we see how symbioses of various kinds have played a fundamental role throughout evolution. As our understanding grows, we will see a greater and more diverse range and influence of such interactive genomic mechanisms for change. Recently, for example, the geneticist Rachel O'Neill, at the University of Connecticut, has proposed that endosymbiotic interactions (chromosome shuffling) between endogenous retroviruses and the genomes of rock wallabies in Queensland, Australia, have given rise over just a few decades to the emergence of eight new species living side by side in the same circumscribed ecosystem. In the words of Andrew Hendry, an evolutionary biologist at the University of Massachusetts, "This is a potentially very interesting mechanism where there isn't necessarily any natural selection driving the differences." How curious that the majority of Darwinians still resist the notion that symbiosis has played an important role in the evolution of complex life forms, most particularly of human biology, behavior, and society. It is my contention that the pluripotent nature of the Genome does not make a special exception for a single species of primate, *Homo sapiens.* Its creative forces are as fundamental to human evolution as to that of any other form of life on Earth.

David H. Janzen, professor of biology at the University of Pennsylvania in Philadelphia, describes mutualisms as "the most omnipresent of any organism-to-organism interaction. All terrestrial higher plants, vertebrates, and arthropods are involved in one diffuse mutualism and many are involved in several." He categorizes these mutualisms into five large groups, the last of which is the sum of all human breeding of animals and crops and other manipulations of nature. The traits resulting from this evolutionary process "are generated largely through the replacement of genetic fitness by the desires of humans, the most mutualistic of all organisms."

· 22 ·

PEOPLE: THE MOST MUTUALISTIC OF ORGANISMS?

> Of course Darwin, like his critics, always believed that natural selection is only one of a number of evolutionary mechanisms; but, whereas they wanted to demote it to a lesser role, he wanted to keep natural selection of small differences as the major cause of evolution.
>
> — MICHAEL RUSE, *The Darwinian Revolution*

OUR HOMINID ANCESTORS first walked upright in Africa about 4 million years ago. The evidence for this can still be seen in Laetoli, Tanzania, southwest of Lake Victoria, where, for a distance of 30 meters, parallel sets of footprints show where a group of three walked side by side. These prehuman ancestors belonged to the genus *Australopithecus,* which means "southern ape." A number of australopithecine species appeared at various times beginning 4 million years ago, one of which, *Australopithecus garhi* (dated to about 2.5 million years ago), had taken the important conceptual step toward becoming the first true human, *Homo erectus,* by inventing stone tools.

"Abel" is the name given to an australopithecine fossil fragment earlier in date than *garhi* that was found in the desert in Chad in 1995 by a group of French paleontologists. Abel is estimated to have lived more than 3 million years ago. All we have left of him is a jawbone containing seven teeth that are deeply scored with transverse lines. These indicate that the creature, thought to have been a young male at the time of his death, suffered recurrent episodes of nutritional deprivation severe enough to stop the growth of his teeth. Malnutrition would also have stunted his bones. Although no more than a fragment of a single individual, Abel demonstrates with brutal clarity the uncertain nature of this creature's life, in particular his lack of access to food, perhaps in difficult climatic conditions such as the dry months of the year.

Australopithecines, such as the famous "Lucy" from Ethiopia, appear to have been very vulnerable creatures. Small of stature and — prior to *garhi* — lacking the skill to manufacture stone weapons, they were easy prey for leopards and other carnivores. Any additional dietary deprivations would have threatened health and even life itself.

Like all living creatures, we humans need to consume sufficient calories from carbohydrates and fats to fulfill our energy needs, and a reliable supply of amino acids, the building blocks of proteins. We also depend on other key nutrients, particularly vitamins. This need, which must have been present from the beginning of our mammalian evolution, caused serious difficulties for all long-distance ocean travelers in the fifteenth century, a period of great European expansionism. Like Magellan's brave crew, the sailors on board their sailing vessels often became sick; on longer journeys, they suffered from a strange and distressing malady that started with swelling and bleeding of the gums and large subcutaneous bruises over their bodies. Uncorrected, this condition worsened into lassitude, coma, and death. Any voyage that extended to several months was likely to be blighted by this mysterious sickness, which outweighed the perils of the deep or the dangers of encountering hostile natives on distant shores. Today we recognize this condition as scurvy, which all too often was the major factor determining the success or failure of the venture.

In 1747, when James Lind, a Scottish naval surgeon, published *A Treatise on Scurvy,* more British sailors were dying from the disease during wartime than were killed in combat. Lind recommended adding fresh citrus fruits and lemon juice to the diets of seamen. But that cure languished unrecognized until Captain James Cook proved its worth during his historic circumnavigation of the globe between

1772 and 1775, when he explored Australia and New Zealand for the British Navy.

The curative factor in fresh fruits and vegetables is now familiar as vitamin C, one of many vitamins we recognize today. In the words of a standard dietetic textbook, these are "organic substances which the body requires in small amounts for its metabolism, yet cannot make for itself at least in sufficient quantity." It is clear that we depend on vitamins for life and health just as much as all animals ultimately depend on the microbes that fix nitrogen or digest cellulose. People who are not well versed in nutrition might not be aware that we also require a number of special amino acids and fatty acids in much the same way. Evolutionary studies have virtually ignored the fact that our health and even our lives depend on these special metabolic entities.

The structures of vitamins and the essential amino acids and fats are no more complex than those of a great many other organic chemicals our body manufactures from the raw ingredients of normal digestion. Why, then, in our long evolution, have we failed to develop the metabolic pathways that would allow us to produce these substances for ourselves? Scientists assume that they must have been readily available from the environment throughout our evolution. But these essential vitamins, amino acids, and fats do not come from the inanimate environment; they derive from other life forms. They would appear to represent missing metabolic pathways, the result of our failure to evolve the genes necessary for their synthesis. Viewed from a Darwinian standpoint, this makes little sense. But viewed from a symbiotic standpoint it would appear to have similar implications to the relationship between insects and their gut microbes: we humans have evolved in a diffuse exosymbiotic relationship with the living providers of these "missing genes." In fact, our human dependency on essential micronutrients produced by other life forms is part of a complex web of such interdependencies in nature. With the exception of the autotrophic bacteria, every life form on Earth has similar exosymbiotic needs.

Today nutritionists recognize fourteen vitamins, derived from both animals and plants. Vitamins A and D come almost exclusively from animal sources: A from dairy products, egg yolk, and oily fish, and D — which is to a limited extent manufactured in skin through the energy of sunlight — mostly from oily fish. The fish, in turn, derive the vitamin largely from plankton living on or near the surface of the oceans, which manufacture it using the energy of sunlight. The two other fat-soluble vitamins, K and E, most commonly come from plants; K is found in fresh green vegetables such as broccoli, lettuce, cabbage, and spinach, while E is found in vegetable oils derived from wheat germ, sunflower seeds, and other plants. The water-soluble vitamins also derive from a mixture of plant and animal sources: vitamin C comes largely from citrus fruits; riboflavin, largely from liver, dairy products, eggs, and some green vegetables. Vitamin B_{12}, which is essential for the manufacture of blood and the health of the nerves and central nervous system, comes exclusively from animals, for example egg yolk and cheese; knowledgeable vegans supplement their diets with B_{12} as well as iron to remain healthy.

One of the most interesting "missing gene" implications is our dependency on the fish-oil-associated omega-3 fats, EPA and DHA. These fatty acids play an important role in fetal brain development and reduce the risk of common illnesses such as heart attacks. A recent study in northern Finland, which gets little sun for half the year, has suggested that dietary supplementation with vitamin D, a component of cod liver oil, in the first year of life may lower the risk of juvenile-onset diabetes.

One has to presume that from long before the three East African australopithecines at Laetoli walked across a field of ash left by a recent volcanic eruption, our early ancestors had access to dietary sources of not just one or even a few of these vitamins and essential fats but all of them. How precarious life must have been for these creatures, particularly in times of want! Many of the serious diseases still prevalent in poorer countries reflect this evolutionary fragility: beriberi is caused by a lack of vitamin B_1 in diets based mainly on refined cereals and lacking in meat; pellagra is caused by deficiency of the water-soluble nicotinamide; and kwashiorkor, recognizable in the tragic newsreel pictures of potbellied children, results from protein deficiency and the lack of essential amino acids.

But even in the most affluent countries, poor or unbalanced diets can result in micronutrient deficiencies. Recently in the United States, the compulsory addition of folic acid to breakfast cereals and whole-grain bread has resulted in a dramatic reduction in the incidence of spina bifida, a serious congenital abnormality affecting the central nervous system. In the words of Margaret Honein, an epidemiologist at the U.S. Centers for Disease Control and Prevention, "The observed decline [in spina bifida] means about 800 more healthy babies are being born in the U.S. each year."

We learned in high school biology lessons that we animals come equipped with a gut, which digests the crude energy, proteins, fats, and minerals that arrive as food, breaking these down into their components, which are then absorbed for energy and body-building purposes. In fact, a significant portion of our digestive tract has little to do with normal digestion: it has evolved in tandem with a veritable zoo of mutualistic microbes. David Janzen notes that in much the same way that plants have mutualistic arrangements with the bacteria, nematodes, fungi, and other "litter decomposers" that surround their roots, animals have mutualistic arrangements with a wide variety of internal microbes. The human large intestine, or colon, is our own internal zoo enclosure. In Janzen's words, "Just as animals can be viewed as frills around a set of gonads, they can also be viewed as frills around a compost heap."

The question, then, is why the vast time span of evolution has burdened an entire kingdom with this nutritional dependency.

In evolutionary terms, the answer is obvious. Cellulose, the tough outer covering of plant cell walls, needs to be broken down if browsing animals are to extract the cells' nutritional contents. The problem was solved for all subsequent plant browsers at the beginning of their evolution by the symbiotic partnerships formed with microbes that had pioneered the land before them. Many of the microbes that now inhabit the intestinal organs of ruminants hopped there from an original independent existence in decaying plant matter. In their present specialized ecology, the microbes derive their own energy needs by fermenting plants ingested by the ruminants. As a byproduct, they produce short-chain fatty acids, which diffuse freely across the gut wall and are grabbed as "food" for the animal's aerobic respiration. This complex symbiosis was exactly the same for the plant-eating dinosaurs as it is for our modern cattle, sheep, and kangaroos.

But take a mental step back and consider the implications. Since all animals must eat plants or the plant-eaters or both, the symbiotic coevolution of herbivores with their microbial zoo must have been a formative step in the evolution of all terrestrial animals.

Consider our human case: every one of us carries twenty times as many living bacteria as we do human cells. They live on our skin, inside our mouth and nose, and on every interface surface inside our body cavities. Why does our immune system not react to them? Because in fact we have coevolved with these minuscule partners. When an unfamiliar bacterium arrives on our skin or enters a body cavity, where it can lead to infection and even blood-borne invasion, it excites a furious attack from our immune defenses. But under normal conditions of health, our immune system does not attack the normal microbial flora, any more than the microbes infect or invade us. Moreover, the numbers and diversity of these microbes are fairly stable — suggesting that we and they have reached some curious yet important equilibrium.

Some 90 percent of the dried weight of our feces consists of the bodies of gut microbes. Although the evolutionary relationship is still poorly understood, our inner zoos appear to be as important to our physiology and health as the warm and nourishing tunnel of our colon is to them. So what, specifically, do these microbes do for us? As Janzen explains, our gut flora — which has very diverse abilities, does its work at low cost to us, and has a considerable capacity to adapt and change when exposed to new foods — "may even be able to do some things that higher animals have never invented." One important benefit of our microbes is that they help to prevent colonization by harmful invaders. Doctors have long been aware that when the normal bowel flora is disturbed, serious bowel disturbance and illness can result.

Susanna Cunningham-Rundles is the director of the Immunology Research Laboratory at New York Weill Cornell Center. One of the pioneers who first defined the AIDS epidemic, she has been looking for ways to treat its complications. One effect of the immunological damage caused by HIV is the stunting of growth in children of HIV-positive mothers. Dr. Cunningham-Rundles was trying to treat an eleven-year-old boy who had intractable diarrhea, mouth ulcers, little appetite, and severe stunting of his growth. When she gave him a fruit drink containing a lactobacillus found in sour milk, the mouth ulcers and diarrhea cleared up and his appetite returned to normal. She went on to treat nine other children with the same condition, five of whom showed dramatic improvements, including significant growth spurts and better immune responses. Dr. Cunningham-Rundles concluded, "The processes of digestion, nutrient metabolism and activation of the immune system are linked in rather surprising ways to the specific composition of microorganisms that comprise the gastrointestinal flora."

One surprising way in which microbes interact with humans is to directly stimulate the immune system. Studies have shown that the passage of beneficial bacteria *through the gut wall* stimulates the production of compounds that play an important part in our immune defenses, including interferon gamma, interleukins, and the gut-associated antibody IgA.

In another demonstration of how our immunity and the intestinal flora may be linked, David Dunne and Anne Cooke of the University of Cambridge have discovered that infestation of the human intestine by parasitic worms may reduce the incidence of autoimmune diseases, such as diabetes and rheumatoid arthritis, in Africans. This may partially explain why in developed countries, where the incidence of parasitic infestation is relatively low, we find a much higher incidence of autoimmune diseases. The Cambridge scientists believe that parasitic worms, such as tapeworms, alter the pattern of immune cell reaction in a way that discourages autoimmunity. The scientists don't plan to infect children with worms, but they are hoping that parasitic extracts might help prevent autoimmune disease in people.

Many doctors and nutritionists have long known that "live" yogurt containing lactobacilli and bifidobacteria can help people suffering severe diarrhea, thrush infestation of the gut or vagina, irritable bowel syndrome, and the general ill health associated with disturbance of the gut flora. Meanwhile, farmers have discovered that similar dietary additives help to promote growth and increase milk and meat yields in their animals.

About half our intake of vitamin K is manufactured by our gut flora, which also releases vitamin B_{12} in a suitable form for absorption through the gut wall. The same flora digest mucus and sloughed cells from the bowel wall to produce substantial amounts of short-chain fatty acids, which contribute up to 10 percent of our energy requirements. Bacteria also play a crucial role in the internal cycle of bile acids, in turn deriving sustenance from the contents of our gut.

Janzen believes that as part of this complex ecosystem of mutualistic interactions, our internal microbes have evolved complex symbioses with each other. Evolutionists such as David Sloan Wilson, William Swenson, and Roberta Elias are looking beyond natural selection of individuals to selection at the level of whole communities of interacting life forms within certain clearly defined ecosystems. On a humble scale, the human gut is exactly such a community. Perhaps it merits a new look from this group selection perspective. In a lecture in Bangkok in February 2002, Joshua Lederberg summed up the complex interaction of life forms that is the living reality of our human nature. "Together with its symbionts/parasites, we should think of each host as a superorganism with the respective genomes yoked into a chimera of sorts." To dramatize this view of the human as a superorganism, comprising the human genome together with scores or even hundreds of ancillary life forms, he suggested the term microbiome to encompass the "entourage of microbial flora with which we share our body space, intracellularly as endosymbionts, but also in our gut lumen, on our skin, mucosal surfaces, and elsewhere."

Of course, the interactions between people and nature extend much deeper and wider than our intestinal flora. Oxygen, put into the atmosphere by photosynthetic plants and microbes, is vital to our evolution and survival. We share the macrosymbiotic oxygen cycle with every animal, plant, fungus, and most of the single-celled creatures, from protoctists to oak trees, and from whales to plankton. Our body chemistry depends on enzymes, many of which depend upon organic sulfur compounds, which in turn derive from the macrosymbiotic sulfur cycle. As Lovelock makes clear, when we die, we return our nitrogen, carbon, sulfur, phosphorus, iron, and even the calorific values of our carbohydrates and fats to smaller creatures in and around the soil, thus still playing a part in the macrosymbiotic cycles of life.

So balanced is the biosphere that if those cycles were interrupted — for example if the microbes in the soil or all the insects in the biosphere died out overnight — all higher plant and animal life, including humanity, would perish. We humans are as dependent on and intrinsically woven into the ecological fabric of our planet as any other life form.

This is an important reminder in these days of overweening self-importance, when we think nothing of exploiting and manipulating nature on a grand scale on land and in the atmosphere and the oceans. A great variety of endangered species depend on our caring about their existence, and in this we touch upon a novel interaction between manipulative humanity and all of life on Earth. Much of Professor Janzen's life and work now involves research in and explanation of how we, the most mutualistic of all animals, need to change the way we think of nature and biodiversity to save what we can of what is left.

We raise animals and cultivate plants, in this way guaranteeing our supplies of food, clothing, and a variety of products, including fertilizer from animal bones. A cynic might point out that we feed and nurture animals only for their usefulness to us, but this is the hard bargain that lies at the heart of mutualistic symbioses. We build hives for bees that in turn provide us with honey. We have selected and harvested the evolved grasses to become our grain crops, and in turn they have provided us not only with the bread of life but also the seeds of civilization. As Jacob Bronowski reminds us, "The largest single step in the ascent of man is the change from nomad to village agriculture. What made that possible? An act of will by men, surely; but with that, a strange and secret act of nature." He goes on to describe how, in the burst of new vegetation at the end of the last Ice Age, a hybrid wheat appeared in the Middle East. Before 8000 B.C., wheat was a poor version of what we see today, one of many wild grasses. But as Bronowski recounts, "By some genetic accident, the wild wheat crossed with a natural goat grass and formed a fertile hybrid."

Today, our molecular biology has taught us the nature of this "genetic accident." The earliest wheat plant cultivated by our ancestors had a poor yield of grain and a diploid genome. An aberration during meiosis resulted in polyploidy and a fleshier kernel, which has been improved through breeding. Through our sentient intelligence, human "determinism" replaced natural selection, and in changing evolution, we became a force of evolution too.

Genetic engineering began in those sunlit fields at the dawn of civilization around ancient cities such as Ur in Mesopotamia. Even in our aesthetic appreciation of beauty, we select and harvest: flowers, flowering shrubs, trees, decorative birds, fish, and so much more that the full list would prove encyclopedic. Today, in a small but significant counterreaction to our own despoliation and destruction, science has begun to make use of the latest genetic advances to store, and even to clone, species at the brink of extinction.

We are quintessentially human not because of our upright posture or our hands and feet but because of our intelligence, which became possible only with the advanced evolution of our sentient brain. Much of that evolution remains a mystery, although a series of advances in understanding have been made in the last two decades. Medical science has, of course, long been aware that the development of the fetal brain requires an extremely efficient blood supply, together with specific vitamins, such as B_{12} and folic acid. Such a dependency would suggest that any novel metabolic pathway affecting key metabolites, perhaps even a more accessible supply of these metabolites, might, through selection and adaptation, lead to evolutionary change in brain growth and development.

For more than a decade, Michael Crawford, a professor at the Institute of Brain Chemistry in London, has been researching the growth of animal brains in relation to the availability of key ingredients in their diets. As he and his colleagues have pointed out, differences in brain power between species do not reflect differences in the metabolic chemistry within the brain but the size of the brain relative to body size. In other words, our intelligence is, through whatever mysterious mechanisms, the function of our relatively large brain. The omega-3 fatty acid DHA is now regarded as essential for full brain development in rats, primates, and the human fetus. There is also abundant evidence that DHA in these species is only poorly manufactured from fatty acids of vegetable origin. The availability of this essential nutrient therefore could have played an important role in the evolution of our human intelligence.

Until recently, paleontologists believed that our early human ancestors were hunter-gatherers on the African savanna. But when Crawford and his colleagues studied the relationship between brain and body size in savanna species, they concluded that as such species evolved larger bodies, relative brain size actually decreased in a logarithmic fashion. The brain of a cebus monkey with a body weight of 2 pounds amounts to 2.3 percent of its body weight; that of a 132-pound chimpanzee, 0.5 percent; of a gorilla, at 242 pounds, only 0.25 percent. In fact the gorilla's brain actually weighs less than the chimpanzee's despite the great differences in their body sizes. At the extreme, a one-ton rhinoceros has less than 0.1 percent of its body weight in its eleven-ounce brain.

The ratio of brain to body size is important. In Crawford's opinion, the biosynthesis of DHA and another essential fatty acid, called arachidonic acid (AA), is relatively slow and may not be able to keep up with body growth, particularly during fetal development in fast-growing animals. A number of experiments have confirmed this. Rats, for example, are better able to convert vegetable precursors to these important fatty acids than guinea pigs, which, in turn, are better at it than wild pigs. Speedy growth of body mass as part of an animal's evolution appears to have been associated with a smaller relative brain size, which paralleled the decline in AA and DHA synthesis. The amount of available DHA and AA appeared to be a limiting factor on the size of the evolving brain.

The question, then, is how did our human ancestors evolve a large brain while living on that same savanna food chain? Judging from the fossil record, the australopithecines showed little increase in brain size throughout the three million years of their evolution. On the other hand, during the one-million-year evolution of *Homo erectus* to *Homo sapiens,* brain size increased greatly. In Crawford's opinion, this growth of the human brain is not explained by Darwinian gradualism. What we see in the fossil record is "a sudden exponential growth of relative brain size in the last 200,000 years or so." Gould and Eldredge might well describe this as a punctuated evolution, accounted for by their variant of Darwinism. But Crawford believes the explanation is quite different.

The earliest true human remains have often been found in lakeshore environments in the East African Rift Valley, while australopithecines are associated with more forested areas. Crawford and his colleagues point out that the dramatic decline of brain capacity in relation to body size of the large mammals and primates that evolved on the savanna is associated with relatively poor access to dietary DHA in that environment. In consequence, the researchers do not believe that *Homo sapiens* could have evolved on the savannas. The richest sources of DHA are the marine and freshwater food chains, and recent fossil evidence indicates that lakeside (lacustrine) and seashore food chains were being extensively exploited at a key time in our evolution from archaic to modern humans. Fish and shellfish would have provided excellent and consistent sources of DHA, which may have stimulated growth and expansion of the brain. The lakeside and coastal environments also offered an excellent source of AA, which is known to stimulate blood vessel growth and development, which is very important for fetal brain growth because of its relevance to the placenta; in humans, 70 percent of the calories passing through the placenta during the critical period of brain development are devoted to brain growth.

Crawford and his colleagues believe that the difference between the relative size of the human brain and that of all other creatures is so large "as to imply that the availabilities of AA and DHA were limiting factors in the evolution of the brain." This hypothesis received some interesting, if indirect, support from a study by Danish doctors reported in the *British Medical Journal* in 2002. Confirming earlier research on the beneficial properties of small quantities of fish oil in pregnancy, the researchers showed that preterm delivery and low birth weights were reduced from 7.1 percent to 1.9 percent if the mothers ate oily fish at least once a week. If Crawford's theory is correct, the implications are far-reaching. A diffuse exosymbiosis between our ancestors and lacustrine and marine life has played a formative role in the evolution of our brain, and of the intelligence that makes us human.

· 23 ·

FROM CUDDLING FISH TO BARTERING CITIES

> But do we have to go to distant worlds to find other kinds of replicators and other, consequent, kinds of evolution? I think that a new kind of replicator has recently emerged on this very planet. It is staring us in the face . . . Already it is achieving evolutionary change at a rate that leaves the old gene panting and far behind.
>
> — RICHARD DAWKINS, *The Selfish Gene*

IN 1970 I WAS PRIVILEGED to be presented with my medical degrees by Sir Kenneth Clark, ex-director of the National Gallery in London and the distinguished author of the book and television series *Civilization.* I still remember his address to the gathered ranks of graduates, impatient to go out and practice their vocation. Decrying the fact that 70 percent of the world's population was still ruled by tyranny, whether of the right or the left, Clark declared, "Aggression and war are the enemies of civilization. The natural desire of people is for happiness, which can only derive from peace and harmony."

How, then, can we equate this ideal view of society with an evolution based solely on aggressive competition? It is not an unreasonable question. The last century saw not only the most lethal wars in history but also some of the worst atrocities. So central has been the perceived role of aggression in history that ethologist Konrad Lorenz and psychologist Erich Fromm, who studied aggression in detail, arrived at very bleak conclusions about human nature. Thankfully, more recent studies have suggested that this pessimism is exaggerated. We humans are far removed from the most aggressive of animals, such as sharks, lions, or tigers. Although our biological and social evolution may predispose us to certain patterns of individual and group aggression, Edward O. Wilson, a neo-Darwinist, concludes that the eventual outcome of our social evolution will be determined by cultural processes, under the control of rational thought. Aggression, as he sees it, may be part of our human evolutionary baggage — predictably involving territoriality, business competition, ethnic prejudices, and, in the sexual sphere, rivalry over partners — but it is only one facet of the complex whole of human behavior. Indeed, there is growing evidence, in the global spread of education, technology, and universal suffrage, for example, that human society is evolving to a more civilized stage in which aggression is being harnessed by the complex subtleties of the democratic ideal.

In a paradoxical way, we appear to be moving closer to the utopian world described by T. H. Huxley, in which "the ideal of the ethical man is to limit his freedom of action to a sphere in which he does not interfere with the freedom of others; he seeks the common weal as much as his own; and, indeed, as an essential part of his own welfare. Peace is both end and means with him; and he founds his life on a more or less complete self-restraint, which is the negation of the unlimited struggle for existence."

Nobody could accuse Huxley of looking at life through rose-tinted spectacles. In his view, this utopian ideal had by no means been realized, if ever it could be realized, because any moral evolution that had taken place had not abolished the deep-seated impulses that propelled "natural man" to follow his nonmoral course. In Huxley's evaluation of late-nineteenth-century industrialized society, the combination of peace and industry was a threat; in making possible the population explosion, it heralded a renewed struggle for existence "as sharp as any that ever went on under the regime of war." Thankfully, his dire forecast did not come to pass. In his prediction of population growth, Huxley was prophetic, but the bulk of the increase was not in the industrialized West, as he had expected, but in developing countries, where poverty and lack of education caused birth control to lag behind. Even in developing countries, population growth has resulted not in a renewed brute struggle between people for existence but in the extinction of other species, coupled with environmental damage to the oceans, terrestrial ecologies such as rain forests, and the atmosphere.

If Darwin were alive today, he would see how the European nations, which had fought wars with each other for more than a thousand years, have moved toward monetary, defense, and possibly even future political union. Meanwhile, the British Empire he so admired has evolved into a commonwealth of nations. We have organizations devoted to global cooperation and mutual support, such as the United Nations, the International Red Cross, and the World Health Organization. Neither Darwin nor Huxley would have dared to hope for today's system of internationally supported tribunals that try war criminals for "crimes against humanity." It is easy for cynics to sneer at such optimism. But however imperfect the present evolutionary stage of such movements, and however flawed and faltering their support in some quarters, it seems nevertheless that human society is evolving toward global democracy and cooperation.

Some might look with skepticism to the recent wars in Africa, the Persian Gulf, and Afghanistan, not to mention the attacks on New York and Washington, as evidence to the contrary. But these events are transitory disturbances in a generally forward trend, triggered mainly by the antidemocratic activities of dictators or the disaffected. Since Lord Clark delivered his speech, a great many nations have moved from dictatorship to democracy, including the relatively bloodless changes seen in Spain and Portugal and, though at the price of a good deal more bloodshed, a great many nations of Eastern Europe, Asia, Africa, and Central and South America.

The question then, from a Darwinian standpoint, is how such a cooperative ideal — however clumsy and still in a process of transition — could be evolving in a species said to be dominated by selfish motivation.

Since the 1970s science has adopted the selfish-gene view of evolution, extrapolating it to biology and ecology in general and to human society in particular. But more recently the Cambridge-based ecologist Lynn Dicks highlighted the same dilemma as Lord Clark. She wrote, "If you accept that evolution is all about selfish genes, the group has no role to play. Survival of the fittest means survival of the fittest DNA. There is no such thing as society. You and I are mere vehicles in which our genes are hitching a lift on the road to posterity. Or maybe not?" With these words, Dicks challenged the aggressive-competitive preoccupations of the twentieth century with the "more caring and sharing" perspective of the opening years of the twenty-first century.

Ras Mohammed is a national park at Marsa Bareika, on the southern tip of the Sinai Peninsula. Here, where the gulfs of Suez and Aqaba come into confluence with the Red Sea, stunningly beautiful coral reefs have carved out a series of localized marine ecologies, circumscribed by barriers of sand. In distinct zones of the reefs known as "cleaning stations," small silvery fish with prominent black horizontal stripes clean the bodies of much larger client fish, including predatorial species that would normally eat them. From May to July 1999, Redouan Bshary, from the University of Cambridge, and Manuela Würth, from the Technical University in Munich, scuba-dived in the area to observe this colorful relationship in detail. The biologists were interested in cooperative behavior that would normally be thought of as limited to primates. Were the cleaner fish, a species of wrasse, capable of social manipulation? Did they engage in preconflict management or postconflict reconciliation?

Individual clients visit cleaners a number of times — sometimes a hundred times — a day to have parasites removed from their skin. The cleaners benefit from feeding on these parasites, each cleaner gorging on as many as 1,200 parasites a day. But analysis of the cleaners' stomach contents has shown that they sometimes cheat by biting off pieces of the client's skin, mucus, and scales. Given the aggressive nature of some of the clients, biting could be dangerous and at the very least result in the loss of future custom. The two biologists were interested in how the cleaner fish coped with this situation. They were bemused to observe the cleaners comforting their injured clients by stroking their dorsal fins or underbellies with their pectoral or pelvic fins.

This manipulative détente was found to be a predictable and reproducible aspect of cleaner fish behavior. And it seemed to work, soothing the clients and seducing them back to the cleaning station. In one observer's words, "Clients obviously 'like' tactile stimulation as they often start drifting motionless in response and they even seek and behave towards an automatically turning brush very much like they do towards a cleaner." Cleaner fish exploited this liking for tactile stimulation and used it to satisfy their own needs. Clients that had been aggressive or uncooperative stopped moving when cuddled in this way and allowed the cleaner to feed off their skins. George S. Losey, a marine biologist at the University of Hawaii in Honolulu, who had investigated other aspects of this behavior, suggested that clients visit cleaners to be stroked rather than to have their parasites removed.

This ingenious symbiosis between the cleaner fish and its clients is not the result of rational thought or planning. It is a perfect example of genomic intelligence, a behavior controlled and inherited through the genomes of both fish. But how did such a cooperative arrangement come about? Its evolution clearly necessitated genomic changes, which Darwinians would explain in terms of selfish genes.

In 1975 Edward O. Wilson published the influential, if controversial, *Sociobiology,* which he defined as "the systematic study of the biological basis of behavior." The book was mainly devoted to animal behavior. Three years later, in *On Human Nature,* Wilson extended his argument to people, contending that "human emotional responses and the more general ethical practices based on them have been programmed to a substantial degree by natural selection over thousands of generations." Finding some commonalities between human behavior and that of the great apes, he acknowledged that at a finer level of classification our species is distinct from Old World primates in ways that can be explained only by the evolution of a "unique set of human genes." The book was greeted by a storm of controversy. Anticipating this, Wilson made clear that he agreed with the geneticist Dobzhansky that human genes have largely surrendered their expression in human evolution to the overwhelming influence of culture. Most, if not all, of our higher human qualities of social behavior, ethics, morality, and aesthetics are learned. But Wilson put a sting in the tail of this argument, emphasizing, like Dobzhansky, that culture itself was ultimately inherited, since it was dependent upon the human genome. And, probing deeper, at a more primitive level of the anatomical brain, he stated that the mind, an "epiphenomenon" of the neuronal machinery of the brain, was the product of natural selection acting on the human genome for hundreds of thousands of years.

To put it another way, every form of life on Earth has predictable behavioral patterns based on a species-specific "intelligence" inherited through its genome, from the simplest of all forms, the viruses, to the most complex, *Homo sapiens.* This concept, which is another way of describing what I have termed "genomic intelligence," is the point Wilson is making. It explains why even simpler life forms, such as bacteria, insects, spiders, and fish, exhibit recognizable lifestyle patterns. And, although Wilson acknowledged the dominance of cultural factors in human social evolution, he did not accept that culture was the whole answer. "The evidence is strong," he continued, "that almost all differences between human societies are based on learning and social conditioning rather than on heredity. And yet perhaps not quite all." He illustrated this with the apparent differences between Chinese American and Caucasian American newborn babies that "cannot reasonably be explained as the result of training or even conditioning within the womb." It would be a gross misrepresentation to call Wilson's proposals racist. Nevertheless, statements such as this provoked an outraged reaction from nonevolutionists and fellow evolutionists alike.

By and large, most evolutionary scientists accept that our physical makeup is the result of our evolutionary history. To state that the Chinese look different from Africans or that Australian aborigines on average have longer legs for their body height than the average Caucasian is not racist. These are nothing more than differences that have arisen in the same basic stock of humanity through evolutionary adaptation to different climates and ecologies. Why, then, did Wilson's extrapolations meet with such acrimony? In fact, some scientists derided Wilson's lack of hard objective evidence for his claims. Others feared that these genetically based views, however innocently intentioned, might be manipulated to promote views of racial superiority, much as Darwin's views were extrapolated out of all proportion by the eugenicists in the first half of the twentieth century.

Many other authors have extended Wilson's genetic determinism to human behavior. For example, Robert Barton and Robin Dunbar, at the universities of Durham and Liverpool, have proposed that the main reason for the growth in human brain size that occurred about 300,000 years ago was selection pressure for social intelligence. In a chapter of the multiauthored volume *Machiavellian Intelligence II,* Barton and Dunbar propose that increasingly complex social interaction was the reason for the evolution of primate intelligence and, in particular, for the expansion of the forebrain that gave rise to *Homo sapiens.* While social interaction surely must have played an important part in human evolution, the Machiavellian argument tends to overemphasize manipulation and deceit at the expense of cooperation. But the core idea — that the "cognitive capability we call intelligence is linked with social living and the problems of complexity it can pose" — seems altogether reasonable in the human context.

We have seen examples of various types of symbioses in evolution, many of which have striking parallels in human behavior; as David Lewis makes clear, where mutualism describes reciprocal benefit between members of different species, cooperation is the equivalent between members of the same species. It seems likely, therefore, that cooperation, embracing many different types of interaction, has played an important role in the evolution of human behavior.

From a selfish-gene perspective, it is difficult to explain why individuals act in unselfish ways. Yet even the most cursory examination of human society, from hunter-gatherers to the global partnerships of the present day, reveals a labyrinth of cooperative interactions. How else do we explain how a religion like Christianity, based on love of one's fellow human beings, grew and spread in spite of the brutal persecution of the Roman Empire? The Christians who accepted martyrdom had no vested interest in the genes of their fellow humans, only faith in their religious belief. When one examines all the great religions, including Judaism, Hinduism, Taoism, and Islam, one discovers a similar level of caring and cooperation in their tenets and commandments. What applies to religions also applies to the great majority of human communities, from tribes to nations.

Just as fish like to be stroked, people enjoy the company and contact of other people. Most of us feel a need to "belong." We derive pleasure from sharing work, conversation, and company with others, quite apart from the obvious intimacies of family and sex. The Internet is merely the latest in a long line of communication possibilities that thrive on our need for social contact, which has been reflected throughout human history in the evolution of language, art, music, and writing. Indeed, the pervasive influence of culture would appear to be an inevitable development of our need for, and love of, communication.

What would happen if cooperation broke down in human society? Few of us grow or raise our own food, nor do we harvest grain or butcher animals; all too often these days, we don't even cook it. Without cooperation there would be no dependable sources of food, supplies of clean water, electrical power for central heating and air conditioning, or the machines we depend on for locomotion. Our manufacturing plants would stop running if we lost the cooperation implicit in the division of labor. Our children would remain uneducated, our sick people would suffer untreated. Without cooperation, there would be no parliamentary democracy, since universal suffrage and the voting franchise are forms of social and political cooperation.

We can be justifiably proud of individual achievement; gifted individuals, such as Einstein, Darwin, Mozart, Picasso, and Mother Teresa, have contributed greatly to civilization, but cooperation is *the* universal feature of human society. As individuals we are undoubtedly selfish, and we have a facility for Machiavellian manipulation, but these traits do not overrule the importance we place on living and working together. Cooperation is, therefore, far too important to be ignored by evolutionary sociology, psychology, or biology. It cannot be dismissed as backward or wishful thinking or mere left-wing ideology. Cooperation lies at the heart of our democratic society, and it embraces the evolutionary root of capitalism itself.

Darwin was well aware of mutual support in the behavior of both animals and humans. In the fourth chapter of *The Descent of Man,* he devotes a great deal of discussion to sociability and group behavior. "Man," he states, "is a social being." Nevertheless, he opted to ignore it in his conceptual framework. We have seen how Peter Kropotkin lost his battle for recognition of cooperation as an evolutionary alternative to unrelenting struggle and pitiless competition. John Maynard Smith admits that for a hundred years after publication of *The Origin,* Darwin's followers completely ignored the role of cooperation, coming to their senses only when a new enlightenment began to permeate the neo-Darwinism of the 1960s. Since then various Darwinians have struggled to explain its role in evolution.

The British biologist J.B.S. Haldane set this ball rolling in the 1930s when he suggested that he would risk his life to save two of his brothers but not one. This approach subsequently became known as "kinship theory," which argues that within families the benefit to an individual who acts "altruistically" can be calculated according to the percentage of genes the individual shares with the person helped. Haldane shared half his genes with each of his two brothers, so only if he were to sacrifice his life to save both brothers would his genes effectively have been saved. In 1964 the late William D. Hamilton worked out the mathematics of risk versus gain in such kin-selected cooperations, and Edward O. Wilson acknowledged his debt to Hamilton in formulating his own theories. Many examples of kin-selected cooperation have been confirmed in the animal kingdom, from social insects to naked African mole rats.

In 1971 Robert Trivers extended the Darwinian interest in cooperation, formulating a theory to explain the frequently observed cooperation between unrelated individuals. Why, he asked himself, would an individual risk his life by going to the rescue of an unrelated drowning man? The explanation, he suggested, was that the two individuals might someday meet again, perhaps on many occasions, and the saver might then reasonably expect some reciprocation of his assistance.

I will leave it to readers to decide whether they are more convinced by that reasoning or by Kropotkin's explanation that when faced with a life-or-death situation, a deeper (in other words, genetically driven) human instinct comes into play. Does anybody seriously believe that a man breaking into a blazing house to save the lives of the screaming children inside does so because he stops to think that some day those children might rescue him? Perhaps he would consider that others might do the same for his children in similar circumstances. But does this mean a childless man is less likely to risk it? Does the boy or the girl who dies trying to save a dog from drowning under the ice on a wintry lake really calculate that, once rescued, the dog might one day pad to his or her rescue? The remarkable, indeed wonderful, fact is that, although they are incapable of any such deliberate reckoning and do not share our selfish genes, dogs have been known to sacrifice themselves for a master or mistress.

It seems more likely that important cooperative behaviors embedded in our human genome — such as love, friendship, and "togetherness" — carry a potential for self-sacrifice in extreme circumstances. Perhaps neo-Darwinians should simply accept that selfish-gene thinking cannot explain everything in evolution. Trivers is altogether more convincing in explaining the importance of "hard bargaining" as an underlying principle of cooperation. Though it lacked mathematical proof, his paper initiated a more thoughtful approach to "behavioral ecology."

A kind of mathematical confirmation of Trivers's theory came a little later, when the American political scientist Robert Axelrod teamed up with William Hamilton. How this came about is interesting.

In the late 1970s Richard Dawkins received an invitation from Axelrod, then at the University of Michigan, to take part in a competition to write a computer program that would play the game "Prisoner's Dilemma." This game strategy, invented in the 1950s by the economists John von Neumann and Oskar Morgenstern, aims to find the optimal strategy an individual might adopt in a conflict situation. It is based on the premise of two prisoners being interrogated separately for a crime they are suspected of having committed together. During his interrogation a prisoner can "cooperate" — in other words, not implicate or blame the other — or he can cheat and turn state's evidence against his partner. The most predictable outcome would be for each to rat on the other and so avoid blame himself. But if they do so, they end up sharing the blame and receive a jail sentence of three years. If only one rats on the other, the cheater gets off scot-free, and the noncheater, who takes the full blame, goes to jail for ten years. However, if both cooperate, the state lacks the evidence to effectively convict them and they each serve a sentence of just one year.

Looked at from each subject's selfish perspective, the best thing to do is to cheat. That way he avoids the danger of a ten-year stretch and at worst serves only three years. On the other hand (and this is the dilemma), if both cooperate, the sentence is only one year. Is there a way out of the dilemma that salvages cooperation? This is the question Axelrod put to those taking part in his competition. The point of the game can be expressed in a more biologically relevant fashion: "Under what conditions will cooperation emerge in a world of egoists without central authority?"

Dawkins thought that the game had interesting potential, but he declined to become a contestant himself. Well aware of the theory of reciprocal altruism, he thought it "an ignoble, low substitute for what might be achieved by conscious foresight in humans." However, he was sufficiently intrigued by the proposal to recommend that Axelrod should approach Hamilton.

A rare rapprochement between neo-Darwinistic and symbiotic theory occurred in Hamilton and Axelrod's search for a solution to the Prisoner's Dilemma. To begin with, the scientists allowed the two individuals to repeatedly practice the game. Through practice there emerged "understandings" or "rules" based on reciprocity. The scientists then invited sixty-two experts, with backgrounds in physics, economics, mathematics, biology, and politics, to design computer-based competitions following the same ground rules. The winning design was the simplest one, a computer program submitted by Anatol

Rapaport, a psychologist at the University of Toronto. In Rapaport's program a player began by cooperating with any new partner. After that first move, the player copied the partner's previous move. If the partner cooperated, the player rewarded him with further cooperation; if the partner cheated, the player punished him in turn by cheating. Grudges were not held. If the cheater reverted to cooperation, the reward aspect kicked in again. This proved to be an extremely powerful mechanism for cooperation.

Axelrod and Hamilton published their findings in the journal *Science,* and Axelrod subsequently expanded the article to a book. Both publications give many examples of mutualism, including the behavioral symbiosis involving cleaner fish. Cheating is possible in behavioral symbioses, and often a relationship that begins as ruthlessly selfish parasitism evolves over time as both partners discover the advantages of bargaining. Cooperation within a species does not assume the parity of gain any more than mutualistic symbiosis between species. It is sufficient that both partners gain some long-term advantage through satisfying each other's needs. However, as Axelrod and Hamilton made clear, long-term stability becomes genetically hard-wired: "Once the genes for cooperation exist, selection will promote strategies that base cooperative behavior on cues in the environment."

I could not better express the rules for mutualistic symbiosis.

Smith and Szathmáry accept that the emergence of modern human society required the cooperation of entities (genes and people) that had been independent and competing. Cooperation, whether between or within species, is not the unusual event one might imagine from the Darwinian preoccupation with competition. Nor is it about Mr. Nice Guy giving away his hard-earned food or labor for nothing. It is about hard bargaining, the staple mechanism of the capitalistic business culture, rather than the enterprise-sapping uniformity of communism. In this synthesis of mutualism and neo-Darwinism, we find no disagreement with Dawkins, even though he is a formidable defender of neo-Darwinism; he writes, "Each one of us is a community of a hundred million million mutually dependent eukaryotic cells." And he rightly goes on to emphasize that every individual animal or plant is a vast inner cooperation ("a community of communities") among interacting layers, like a rain forest.

In analyzing the "coinage" of mutualistic symbioses between species, Janzen arrives at conclusions that are equally relevant to cooperation between individual members of the same species. The core of a mutualism is the payment of what is "fitness-cheap" to you in exchange for what is "fitness-dear" to you. What is easy for an ant, its mobility and aggression in protecting the acacia tree, is impossible for the tree. What is easy for the tree, affording a shelter within its hollowed-out thorns, or diverting a tiny fraction of its energy to providing more edible extrafloral nectaries or leaflet tips as food, is impossible for the ant.

From the Darwinian perspective, an important question remains: if cooperation is a powerful evolutionary mechanism, then how, given selfishness at the levels of the individual and the individual gene, is cooperation maintained and inherited at a genomic level? The answer is that selection operates at different levels.

Even at the level of genes within a genome, some genes behave selfishly, such as rival versions of the same gene and "jumping genes," which replicate out of phase with the chromosomes as a whole. Jumping genes have been known to spread rapidly throughout a population, inserting far more than a single copy of themselves into a genome, sometimes as much as half of the total DNA. How, then, does the genome cope with selfish behavior that threatens its fitness as a whole? Smith explains that if the selfish behavior of an individual gene causes a loss of reproductive fitness, group selection at the chromosome or whole-genome level will tend to suppress that behavior.

Ernst Mayr strongly believes that selection also acts at the level of the organism as a whole. The death of any life form means the death of all its component cells, just as the death of a genome results in the death of its component genes. An elephant, for example, depends on cooperation between different kinds of cells. These cells all share the same genes, so the cells throughout the elephant have a survival interest in cooperation. If one cell behaves selfishly in producing far too many copies of itself, a cancer results. This selfish manifestation destroys the tissues or organs in which it grows and ultimately causes the death of the whole individual. Therefore the cells making up a plant or animal have a survival interest in cooperation, with selection operating at the level of the plant or animal as a whole.

Darwin first suggested that selection at the group rather than the individual level might explain mutual support. A group of people who are kind and helpful to each other might not fare especially well individually, but as a group they might succeed better than other groups, so the tendency to be kind and helpful will spread throughout the population. Smith and Szathmáry advance a similar group-selection argument for the evolution of symbiosis. The million-dollar question is whether such an argument can be extrapolated to cooperation as an evolutionary adaptation for a whole species, and in particular to what we see in human society.

We are just beginning to understand some of the genetic, and thus evolutionary, aspects of complex evolved behaviors. Humans have the largest forebrain in relation to body size of all animals, the prefrontal region amounting to more than 20 percent of the total brain weight. This massive enlargement of the prefrontal region, even when compared to our primate cousins, is so marked and specific that some biologists consider its evolution as that of a specialized organ. It plays an important role in higher social functions, including the morality and ethics that underpin socially acceptable behavior.

Recent forensic evidence has pointed to a relatively high incidence of damage to or impairment of the prefrontal region in "aggressive antisocial" behavior, including serial rape, murder, and, most notably, murder followed by cannibalism of the victim. Other research has confirmed what neurobiologists have always suspected: that specific pathways in the more primitive regions of our brain affect other human behavioral patterns linked to our emotions. One of these pathways is the "reward limb" of the mesolimbic system, which works through the mediation of the chemical neurotransmitter dopamine. A release of dopamine in this system leads to feelings of pleasure or elation. Release of another neurotransmitter, noradrenaline, is associated with improved alertness, concentration and memory. Another specific center, known as the locus coeruleus in the upper midbrain, is believed to play a part in the psychological registration of complex sensations and emotions, including pain, unhappiness, and loss. These pathways have links to the sentience-associated lobes in the brain as well as to sensory areas such as vision and smell. They also link to such visceral functions as adrenaline release, heartbeat, skin pallor, and sweating — in fact the physical symptoms we typically associate with powerful emotion.

Although our understanding of these pathways remains sketchy, they afford at least a partial anatomical and physiological basis for how specific human experiences and behavior, such as the joy of love, sharing, cooperation — and also the visceral pleasures of competition and aggression, and even the experiences of war — can be linked to primal brain pathways, which are part and parcel of our evolutionary inheritance.

In recent years a new generation of evolutionary psychologists has extended the ideas of social Darwinism to the highly charged arenas of sexual behavior and misbehavior. Merging neo-Darwinism with "cognitive psychology," some have proposed that differences in gender behavior are hard-wired into our genome because males and females have followed different evolutionary trajectories. This focus on an arena in which many of society's anxieties and concerns reside worries many scientists. For example, Stephen Rose, an evolutionary biologist at the British Open University, has accused the "sexual Darwinists" of having a "Flintstones' view" of the human past, based on mere speculation. In the opinion of British journalist Dave Hill, writing in *The Observer,* the sexual focus is also disturbing to sociologists, who feel that evolutionary psychology lends momentum and a veneer of intellectual validity to the media's obsession with the "sex wars." The sociologists' opposition is hardly surprising when Hill explains what the evolutionary psychologists are proposing: "that gender differences, and even such diverse aspects of behavior as infidelity, jealousy and infanticide, can best be understood not in terms of education, learning and conditioning, but as the consequences of evolutionary pressures that took place in our distant past." Recently Randy Thornhill, a biologist at the University of New Mexico and his anthropologist colleague Craig Palmer, have extrapolated from insect and primate behavior to claim that "rape . . . should be viewed as a natural, biological phenomenon that is a product of the human evolutionary heritage."

David Buss, a professor of psychology at the University of Michigan, helped set this highly controversial debate going in 1994 with the publication of *The Evolution of Desire*. In this book he asked, "Why do people devote so much time, passion and energy to mating?" Mating, marriage, rearing and caring for a family: all suggest the vital role of cooperation. But in setting out to explain the pervasiveness of love in all cultures, with its key components of commitment, tenderness, and passion, Buss takes an exclusively neo-Darwinian perspective. A reader hoping for a single index reference to symbiosis, cooperation, or mutualism will be disappointed, but there are entries for competition, aggression, conflict, divorce, incest, mother-son incest, rape, promiscuity, homicide, and even the contribution of *Playboy*.

This unbalanced approach to evolutionary psychology is all the more worrying because it has become one of the most rapidly growing areas in debate about evolution. And because these topics are relevant to everyday society, we need to have a broader, more balanced perspective on them. Only then will we gain a fuller understanding of the complex weave of human evolution, in which cooperation is as important as competition.

Before the 1960s, certain evolutionists, in particular Warder Clyde Allee at the University of Chicago, had pioneered the study of cooperative behavior "in everything from worms to humans." Allee and his colleagues believed that animals showed two types of social interactions: one involved a self-centered egotistical drive; the other, group-centered drives leading to the preservation of the group as a whole. Allee also believed that selection at the level of the group was, if anything, more important than that at the individual level in shaping the evolution of behavior. After Allee, the baton for group selection was taken up by an ornithologist, V. C. Wynne-Edwards, who proposed that the chorusing behavior of songbirds was a means of censuring their own groups to avoid overpopulation and possible extinction.

The reductionists, including George C. Williams of Princeton and Richard Dawkins, fought back from the 1960s onward, arguing that we could explain everything more simply on the basis of individual selection and there was no need to involve complicated entities such as groups.

This opposition halted the group selectionist approach to behavior for three decades. But a new school of thought rose from the ashes. David Sloan Wilson, a professor of biology at the State University of New York at Binghamton, and Elliott Sober, a professor of philosophy at the University of Wisconsin, used a more mathematical approach and included detailed genetic models of how group selection might work. In particular they identified two opposing forces: selection for and against cooperators both within and between groups. Within groups, as the individual selectionists had long argued, cooperators lost out. But between groups, those whose members cooperated were more likely to win out against those whose members were selfish. In other words, cooperation became an evolutionary force when selection between groups was more important than selection between individual members within a group.

This kind of group selection may explain the evolution of the social insects, such as honeybees, into superorganisms in which the entire colony behaves as an integrated whole.

Group selection may also explain some aspects of the evolution of genomes and even whole multicellular organisms, with their need for cooperation between individually selfish genes and different tissue cells. The relevance to human society, with its villages, towns, cities, and states, becomes obvious. The religious sect of Hutterites, now primarily based in Alberta and Saskatchewan, whose individual members eschew ownership of private goods and property, work together for the common good. In traditional Hutterite society, cheaters are banished. So efficient are their arrangements that the group is remarkably successful in cultivating marginal land. But before rushing off to join such a group, one should consider that the individual contributions of an Albert Einstein or a Ludwig van Beethoven would never have emerged in such a society.

Whatever one's views on group selection, there can be no doubt that cooperation has major societal and political implications. Dawkins wholeheartedly supports Axelrod's recommendation of the benefits of cooperation to U.S. senators, and, even further, to the uneasy balances and alliances between friendly and not-so-friendly states.

Edward O. Wilson believes that 95 percent of our societal evolution took place when people lived in the extended family formation of the hunter-gatherers. By the time *Homo erectus* arrived, more than a million years ago, the increasing size of the human brain led to babies being born that were in essence still gestating. The anthropologist Owen Lovejoy, of Kent State University, believes this had further implications, suggesting that humans adopted an upright posture because it freed the arms to carry such dependent infants, as well as useful items such as food. "A relatively bizarre form of locomotion, bipedality, only makes sense as part of a larger adaptive breakthrough in the entire biology of this particular hominid."

The ability to carry infants and food must have been an important step in social and sexual evolution. If, as Lovejoy explains, one looks at the female of our closest primate relative, the chimpanzee, "her reproductive pattern is pretty precarious. As among other apes, male chimps are only interested when the female is in estrus. Otherwise, there's no pair-bonding, no real cooperation between male and female in matters of food-finding or raising infants." The female chimp has to do everything by herself. The young are obliged to cling for life to the mother's fur as she searches for food, expending so much time and energy that she cannot afford to have more than a few offspring in her lifetime. The mean length of time between births among chimps is five years, with an inevitably high incidence of infant mortality.

Those wonderful Laetoli footprints are more informative than a thousand scientific papers about our human history and evolution some 3 million years ago. It seems likely that they record the passage of a family of australopithecines. If true, this would suggest that the family stage of our social evolution was already established at this "prehuman" stage, long before we had invented stone tools. Another important observation is that they walked upright on feet that were no longer prehensile, like those of their arboreal ancestors; their feet were almost the same as our own today. Since a primate, such as a chimpanzee, can run faster on four limbs than we can on two, clearly our hominid ancestors must have gained some alternative advantages from this change to an upright posture. As omnivores, they must have adopted a search-intensive strategy to find food, a lifestyle that would have been more difficult still for the female with offspring to look after. The ability to walk while carrying an infant has obvious advantages: it becomes an adjunct to the evolution of social cooperation.

The gravid mother, particularly in the later stages of pregnancy, moved more slowly and was thus even more vulnerable. Her ability to survive was enhanced by the protection and food-providing advantages of a group. The growing infant was more vulnerable still, and the mother, needing to carry the infant with her as she traveled, continued to benefit from the protection of being in a group. Most evolutionary sociologists agree that the family unit, and its expansion into the extended family, must have provided the beginnings of socially cooperative groups in humans.

As Lovejoy notes, male apes that do not bond with females develop fighting canine teeth, which they use in a variety of sexually competitive behaviors. But the evidence suggests that the australopithecine male was not sexually aggressive in this way, since his canines, as in all subsequent humans, are similar in size to the female's. A male that fought off other males to impregnate a female and then abandoned her might have been able to pass on his genes, but the offspring of a male that cooperated with his female partner, protecting her and their child and providing them with food, would have been more likely to survive and thus pass on the paternal genes. This cooperative bond, with its implied monogamy, might also help to explain the disappearance of estrus in female hominids. Regular sex between partners repeatedly sealed the bond while guaranteeing the cooperative male that the children of the partnership were his. In Lovejoy's words, "The more cooperative males became favored — by females and by natural selection."

It is all very well to hypothesize in this way, but evolutionary science rightly demands proof. A recent study of twenty-four present-day hunter-gatherer societies on four continents sought to identify universal behavioral patterns that would reflect the evolution of such groups. The study consistently produced evidence of egalitarian behavior leading to deep-seated cooperation as a central and consistent feature. Peter Dickens, a senior research fellow at the University of Cambridge, has attempted a modern reappraisal of the role of social Darwinism. Exploring our egalitarian predispositions, he describes how humans might have adapted in ways "that reduced conflict and maximized social harmony in the interests of survival." Sharing food was one way to reduce conflict and increase harmony. "An evolved collaboration in resource use can be seen as simply a very effective response to the problem of survival and reproduction."

Call it collaboration, cooperation, or simple barter — it does not matter! All of these strategies within a species have much the same mechanisms and implications as mutualistic symbiosis between species. Any psychologist or sociologist trying to explain human society must take into account that we are an intensely interactive and cooperative species. Barter is part of the very nature of human society, as it has been from the evolution of the first hunter-gatherers, to the simplest villages, to the great metropolises that are seen as the beginnings of modern civilization.

In the mid-1990s, the importance of trade received unexpected support from discoveries made in Peru by Ruth Shady, an archaeologist at the University of San Marcos in Lima. In 1994 she began to excavate the ruins of six large pyramids in a desert location twenty miles inland from the Pacific coast. The pyramids were found to mark the site of a city, Caral, which was carbon-dated to almost 5,000 years old, by far the oldest city found anywhere in the Americas. There was no evidence of the working of metal; Caral was a Stone Age metropolis. Many archaeologists saw it as a prime site in the world for evaluating how civilization first came about.

The ruined city was so vast and complex, Shady called on Jonathan Haas from the Field Museum in Chicago, and his wife, Winifred Creamer of Northern Illinois University, both of whom had long been interested in the origins of civilization. The leading proponent of conflict as the central force behind the origins of cities, Haas had found aggression-based mechanisms wherever he looked, from the cities of Central America to Egypt, Mesopotamia, India, and China. He had advanced the theory that fear of conflict had forced small groups to come together for mutual protection. This had given rise to the first cities, and with them the beginnings of our modern civilization. But when Haas searched for evidence of warfare in Caral, he did not find it. There were no surrounding walls or defensive fortifications, no stone cudgels or battle-axes, no carved depictions of warriors fighting or of the torture of captured enemies so commonly found on the walls of later cities. Where in other cultures the remains of children had all too often proved to be evidence of human sacrifice, in Caral the remains of children buried under the floors of the grandest houses showed obvious signs of caring and tenderness.

What the archaeologists did find was evidence of trade on a grand scale. The local diet was based on fish, which had to be brought into the city in large and regular quantities from the coast. Other commodities were clearly brought in from the Andean rain forest, 200 miles away, or from trading over considerable distances up and down the coast. The basis of exchange was cotton grown in fields around the city, made possible by irrigation channels fed by the local rivers.

Among the archaeological finds were imported condor bones that had been made into elaborately carved flutes, evidence of the importance of music and dance, while coca leaves suggested a liking for stimulants and aphrodisiacs. In some ways it seems that cities have not changed much in 5,000 years. All the evidence pointed to a Stone

Age civilization lasting at least 1,000 years that knew not only how to feed and clothe its citizens but also how to enjoy itself without the horrors of conflict. For Haas the conclusion was inescapable: "You have to change your whole mindset about the role of warfare in these societies. It looks like exchange is now emerging as the most effective theory we have today to explain how this system developed."

· 24 ·

THE WEAVE OF LIFE

> Millennia of experience show that by entering into a symbiotic relationship with nature, humankind can invent and generate futures not predictable from the deterministic order of things, and thus can engage in a continuous process of creation.
>
> — René Dubos

In a letter to Thomas Poole dated March 23, 1801, the poet Samuel Taylor Coleridge expressed his belief that "the souls of five hundred Sir Isaac Newtons would go to the making of a Shakespeare or a Milton." One can sympathize with his viewpoint, for without a sense of beauty, objective examination reduces the wonder of life to mundanity. But the search for objective truth does not necessarily imply the loss of aesthetic appreciation, as Albert Einstein explained in his credo before the League of Human Rights in 1932:

> The most beautiful and deepest experience a man can have is the sense of the mysterious. It is the underlying principle of religion as well as all serious endeavor in art and science. He who never had this experience seems to me, if not dead, then at least blind . . . To me it suffices to wonder at these secrets and to attempt humbly to grasp with my mind a mere image of the lofty structure of all there is.

Through probing the secrets of nature we increase our appreciation of it by adding intellectual depth to our instinctive empathy. And at the very heart of our wonder about the origins of life and its diversification into the remarkable and beautiful weave of our Earth today lies a sense of delight. Life, in its complex, almost unbelievable journey from such humble beginnings, becomes all the more beautiful, an even greater object of admiration.

When Charles Darwin published his theory of the origin of species by natural selection in 1859, he changed the way we look at life. But only some of the strands of Darwin's theory, as teased out by Mayr, have stood the most rigorous test of validity over time. His first strand — that evolution is the explanation of the origins of life — is accepted by most educated people, whether or not they regard themselves as scientists. Darwin shared the discovery of evolution with many others, the most influential of whom was Lamarck, but also including Alfred Russel Wallace and his grandfather Erasmus Darwin. However, as Louis Pasteur made clear, in science the credit goes to the man or woman who convinces the world rather than to the person who first thinks of the idea. It is, therefore, to Darwin's great credit that he was the man who convinced the world of the truth of evolution.

The second of Darwin's five strands is that all of life arose through common descent from a simple ancestor. This has been confirmed not only by the fossil record but by advances in our understanding of molecular biology. The decoding of DNA and its application to Mendelian genetics culminates in the elucidation of our own human genome, with its many commonalities and contributions from simpler creatures. It is unlikely that life began as a single creative spark enlivening one small pond. Our modern understanding of self-replicating strings, and of the importance of gene transfer across the entire bacterial world, suggests multiple sources of origin, with a vast proliferation of swapping and exchange. But, given the genomic commonalities of the five taxonomic kingdoms, the conceptual difference is minor, and once again Darwin must take full credit.

His third concept, of new species arising from old and, over geological time, giving rise to the great diversity of life, has also been amply confirmed. But his fourth and fifth concepts — the slow accumulation of gradual change under the creative influence of natural selection — offer an incomplete explanation of the real complexity of evolution. Darwin did not realize that the interactions between different species which we know as symbioses are important forces for evolutionary change. Moreover, they have the potential of giving rise to sudden and radical changes, the saltations that he vehemently denied were part of how natural selection worked.

Even after de Bary's coinage of the term "symbiosis," nineteen years after publication of *The Origin,* most biologists thought the phenomenon nothing more than a curious observation. An inspired few suspected its enormous potential, but evolutionary biology needed the enlightenment of modern molecular biology, with its understanding of genetics and genomics, before it could fully comprehend the spectacular evolutionary force of symbiosis.

Today leading Darwinians such as Richard Dawkins acknowledge the creative role of symbiosis in evolution. John Maynard Smith and Eörs Szathmáry have accepted its role in at least three of their five major transitions of life. I have introduced evidence from fields as diverse as molecular biology, nutritional science, genomics, and the epidemiology of epidemic infections, to help explain the many capacities for evolutionary change implicit in the pluripotent wonder I have labeled the Genome. The role of the Genome cannot be downplayed: it has initiated every step in evolution since the time of the first self-replicating molecules. Moreover, because it embraces all mechanisms of hereditary change, the Genome concept provides the variation needed for Darwinian competition as well as for diverse interactions of exosymbiosis and endosymbiosis, all forms of horizontal gene transfer, and also the genetic changes that underlie cooperation within a species, including our own.

Darwin's main innovation, the creative force of natural selection, has withstood all challenges to its role in Darwinian theory. But when it comes to symbiosis, natural selection has shortcomings. Even in the least intimate of exosymbioses, such as that between the hermit crab and the anemone, the partnership arises between organisms with established capabilities, so natural selection is limited to fine-tuning after the creative partnership has already formed. As Law suggests, once the symbiosis is established, each partner is likely to show resistance to further selection pressures because individual change could upset the equilibrium of the symbiotic whole. Indeed, Smith has proposed that in symbioses of this nature, selection probably operates at the level of the partnership. And the creative mechanism of the lightning strike of endosymbiosis is not natural selection but the act of union of two or more existing genomes.

I began this book by describing Kwang Jeon's witnessing of the emergence of a new life form, the xD amoeba, from the endosymbiotic union of *Amoeba proteus* and the X-bacterium. As this singular experiment showed, *genomic forces* came into play almost immediately after the bacteria entered the amoeba's cytoplasm, changing the genetic makeup of the individual symbionts so radically that neither could survive any longer as an individual organism. But there was a significant problem with this great evolutionary innovation. As Jeon took pains to point out to me: "Here is the paradox. We cannot see any survival advantage of the symbiosis. Even today, the xD amoebae are not quite on a par, in terms of survival ability, with their former noninfected equivalents." As stated earlier, the hybrid life form might not have survived in nature, because the first generations were extremely vulnerable. But this does not mean that *a* hybrid could *not* have emerged in the real world.

In nature, a wide range of symbiotic encounters would surely take place between different strains of the bacterium and amoeba, each giving rise to a different hybrid organism. Sooner or later, a hybrid might well emerge that was equipped to survive and prosper. The mechanism of its evolution would still be endosymbiotic experiment. But the new life form would have to be able to compete with other species, including the parent strains of amoebae, in that ecosystem. And this is the stage at which natural selection takes on the more limited role of further honing the new symbiotic life form.

One possible mechanism for survival would be the transfer from the bacterium to the hybrid of some pre-evolved ability that would give it a competitive edge over the parent stock of amoebae. Aggressive symbiosis — as we have seen in the examples of the luminous bacteria that paralyze the prey of certain worms, the virus that partners the ichneumon wasp, and the Clavicipitales fungi in their symbioses with grasses — is one obvious possibility. In fact, many types of amoebae and other protists harbor endosymbiotic bacteria that contribute to the partnership in a variety of ways. One spectacular example is the giant Archaeprotist genus *Pelomyxa palustris,* which, although it has no mitochondria, can utilize oxygen at lower concentrations than eukaryotes normally require. Its mitochondrial equivalents are three different bacterial symbionts, at least two of which produce methane gas instead of carbon dioxide.

But all such speculations aside, Jeon's discovery is an example of the ultimate truth as far as evolution is concerned: the creativity of the Genome, a distinct and separate force for change, must work hand in hand, in greater or lesser degree, with the editorial force of natural selection. This is such a key to understanding that the whole foundation of evolution can be grasped in a sentence.

Since the 1960s, evolutionary scientists have come to redefine Darwinism in an exclusively reductionist way. But Ernst Mayr, unlike many of his contemporaries, does not see the reductionist school as wholly representative. He believes that Darwinism is "heterogeneous," with a wide diversity of thinking among its followers. Mayr bemoans the fact that even today Darwin's theory is spoken of as if it were a singular entity rather than a heterogeneous collection of ideas. The reductionist focus of neo-Darwinism over the last three decades has obscured the fact that from the start markedly different interpretations of Darwinism have arisen. But no Darwinian theory could possibly cover all of evolution. In fact, the terms Mayr uses to describe Darwinism could be applied to evolutionary biology as a whole, as no longer a simple theory that can be proved true or false but as a broad scientific movement embracing a variety of interpretations, "a highly complex research program that is being continuously modified and improved."

At last a consensus of rival opinions is possible. Werner Schwemmler, a biology professor at the University of Berlin, has suggested a balance to bring about a new level of understanding: "The combination of the two explanations (Darwinian gradualism and symbiotic saltation) makes further progress toward a unified theory of evolution possible." In Schwemmler's conceptual framework, known as endocytobiology, the biological study of all intracellular relationships is viewed as a discipline in itself. This widening of perspective has exciting potential both for evolutionary biology and for its extension to philosophy. Darwinism has long served as a rallying banner against the attacks of the creationists. The time has surely arrived when we must move on. As David Lewis has urged in defining mutualism, a new generation of evolutionary scientists should put aside the semantics and focus on the scientific mechanisms of change.

On February 12, 2001, when the first analysis of the human genetic code was published, it became clear just how closely related we humans are to one another. Every human being on Earth shares 99.99 percent of his or her genetic sequences with everybody else, regardless of race, ethnicity, or religion. Differences in skin pigment, hair color, body proportions, and facial appearance are nothing more than regional adaptations to climate, vitamin D availability, and other factors. Darwin would have been genuinely astonished to learn that, as modern genetic studies have revealed, individuals often have less in common genetically with their neighbors than with others from an entirely different ethnic group half a world away.

In the words of Craig Venter, formerly of Celera Genomics, the privately funded organization that jointly worked out the human genome, "No serious scholar in this field [genome research] thinks that race is a scientific concept."

In his entertaining and provocative book on cooperation, *Cheating Monkeys and Citizen Bees,* Lee Dugatkin confessed that although he is a firm believer in Darwinian evolution, "my Judeo-Christian heritage, as well as my strong belief in one God who holds people accountable for their actions, affects every aspect of my life." This religious perspective so colors his view that he believes animals have "no inherent worth" except in relationship to humans. I do not share that view.

While I cannot pretend to consider animal or plant species as important as my own (my genes are too selfish for that), I do respect the inherent worth of biodiversity and I see a concern for the survival of endangered species as entirely logical. I know that in this wish to protect, as far as is humanly possible, Earth's living inheritance, I am one in spirit with James Lovelock, who sees in biodiversity a vitally important aspect of the georegulatory aspects of Gaia.

To feed and shelter ourselves, we humans have destroyed great areas of wilderness, including rain forests and coral reefs, and we have polluted the oceans and atmosphere, probably speeding up global warming. Yet on this same universal scale, we have shown a deep and abiding concern for our environment. During times of stress, such as war, we have shown ourselves capable of terrible brutality to our own species; to deny the reality of our cultures of war, selfishness, and aggression would be foolish. But these forces are counterbalanced by our curiosity and caring about other people and nature, and especially by the most cherished human quality of all, the one that, along with sentience, defines us as human: our capacity to "love."

In classical mythology, there is a story about the Lydian maiden Arachne, who so excelled at the art of weaving that she challenged Athene to compete with her in creating a tapestry that would best portray the dignity of the Olympian gods. Athene's fury at Arachne's skill drove the maiden to attempt suicide, prompting the goddess to save her life by the cruel whim of turning her into a spider, forever condemned to weave cobwebs instead of beautiful tapestries.

In evolutionary science, humanity has also challenged this Olympian hubris. But this tapestry is not woven on a loom; rather it is born in the vastness of space, after the colossal explosion that brought into being all matter, all laws, all space and time. From one of the billions of dark Magellanic Clouds, the descendants of the colossal supernovae, a small galaxy is born, centered on a very ordinary star. Around it planets aggregate, one of which is the molten ball of rock that is our nascent Earth, an oblate spheroid of rock and magma, animated by tectonic forces.

Here the weaver begins, with her shuttle powered by the electromagnetic energy from that early sun. As the shuttle starts to create its pattern, mountains rise and fall through the dynamic of Earth's internal forces and oceans move restlessly under the influence of a coevolving moon. The first landmasses emerge under the spiraling mists of the early atmosphere. Most wonderful of all is the weaver herself. This genomic Arachne, whose home will become the living cells of her own creation, needs neither fingers to weave nor a plan to work from. She is the most mysterious of entities, the pluripotent source of creativity itself. Her song, accompanying the creation of the ever more complex tapestry, is not the single monotonous note of neo-Darwinian mutation but a symphony of all possible notes, voices, and instruments. In this glorious medley, and in cooperation with the inanimate world of nature, a single all-seeing eye, devoid of human morality or sentiment, views the developing weave and decides yes or no or tries something new for a while. As the weaver unpicks and refines the evolving tapestry, its colors and forms merge inextricably with one another, interacting locally and globally with the raw substance of the mountains and atmosphere, oceans and crust. What emerges is every terror and wonder of life.

The skeptical Coleridge was mistaken. The greatness of Isaac Newton is not diminished by the artistry of Shakespeare or Milton. On the contrary, great science adds an extra dimension of understanding to the existing aesthetic sensibility, so that in the numinous whole, we experience what Einstein meant by his sense of the mysterious.

· *Notes* ·

Introduction. A Mystery of Nature

2 All Jeon quotes are from my interview with Professor Jeon.

2 "Chance favors": *'Dans les champs de l'observation, le hasard ne favorise que les esprits préparés.'* From Pasteur's inaugural lecture at the University of Lille, Dec. 7, 1854.

Chapter 1. The Origins of Life

13 "the book that shook the world": Mayr 1991: 7.

14 Darwin kept changing his mind: ibid.: 21.

14 "Natural selection is not the wind": Moore 1979: 316.

Chapter 2. The Other Force of Evolution

15 The crab and the anemone appear to recognize each other: Thomas 1974: 40.

17 concept of genomic intelligence: Ryan 1997.

19 "a neutral term was required": Sapp 1994: 6.

19 de Bary is now acknowledged: ibid.: 7.

19 "animals containing chlorophyll": P. Geddes, "Further Researches on Animals Containing Chlorophyll," *Nature* (Jan. 26 1882): 303–5.

20 "What is the physiological relationship . . . ?": ibid.: 304.

21 two subsequent books: Geddes and Thomson 1889; 1912. See also the broad discussion of this period in the history of symbiology in Sapp 1994.

21 "Whether true truffles also establish a mycelial connection": B. Frank, "On the Root-symbiosis-depending Nutrition through Hypogenous Fungi of Certain Trees," *Berichte der deutschen botanischen gesellschaft* 3 (1885): 128–45 (trans. J. M. Trappe), published in *Proceedings of the Sixth North American Conference on Mycorrhizae* (June 25–29, 1984). I am deeply grateful to David Read, of the botany department of the University of Sheffield, for much of this information on Frank and mycorrhizae.

Chapter 3. From Anarchy to Cooperation

25 "A month ago I was invited": Kropotkin 1902: xix.

26 "We looked vainly for the keen competition": ibid.: Introduction.

27 "This suggestion seemed to me so correct": ibid.: xi.

27 Huxley's essay: Huxley 1888. A copy of the original essay can be obtained from the British Library. It is also reproduced in full in the 1989 edition of Kropotkin's book.

28 "For thousands and thousands of years": Kropotkin 1902: chap. 1.

28 the Neandertals were more apelike than human: Stringer and Gamble 1994. There is an exceptionally good Web site on the Neandertals at http://sapphire.indstate.edu/~ramanank/.

29 "I conceived since then serious doubts": Kropotkin 1902: Introduction.

30 Engels . . . dismissed Malthus's theory: Singer 1999: 26.

30 Wallace . . . pointed out that Malthus had ignored: Dickens 2000: 10.

31 Kropotkin was unable to see that Darwinians: Dugatkin 1999: 32–34.

Chapter 4. The Price of Failure

32 The Leonard Darwin quote is from G. K. Chesterton, *Eugenics and Other Evils* (1922).

33 most of what is unusual about man: Dawkins 1976: 192.

34 "I use the term Struggle for Existence": Darwin 1859: chap. 3.

35 their role was essentially reproductive: Dickens 2000: 21.

35 "So often did the opponents of regulation appeal": Singer 1999: 12.

36 "Since they proclaim what ought . . . to be done": Carneiro 1967: xliv.

37 "We civilized men . . . do our utmost to check": Darwin 1871: chap. 4.

37 "Man is more courageous": ibid.: chap. 19.

37 the example of Beatrix Potter: Sapp 1994: 5; Desmond and Moore 1992: 470.

38 "The careless, squalid, unaspiring Irishman": Darwin 1871: chap. 4.

39 Blacks were stupid, inferior, and lazy: ibid.: chap. 7.

39 if these . . . "do not prevent the reckless": ibid.: chap. 5.

39 "It needs no argument to prove": Huxley 1888.

40 "There is, perhaps, no more hopeful sign": Huxley 1897: 51.

40 Indeed, it was readily accepted: Desmond and Moore 1992: 579.

40 In 1883 he took these theories . . . further: Galton 1883: 24–25.

41 Gotto . . . founded the Eugenics Education Society: Mazumdar 1992: Introduction.

42 concluded "that if the prolific breeding . . .": ibid.: 3.

42 Fisher . . . was tainted: Ruse 1999: 84.

42 tuberculosis is . . . an infectious disease: see Ryan 1993.

43 Inge expressed his fears . . . Punnett . . . issued a call: Mazumdar 1992: 97–101.

44 first International Congress of Eugenics: ibid.: 90.

44 social Darwinism became even more firmly entrenched: Ryan 2001.

46 "It is the struggle for existence that produces": Hawkins 1997: 274.

46 eugenicist . . . Lapouge, whose proposals contained: Lapouge quoted in Schneider 1990: 61.

46 *La Société Pure:* Pichot 2000. See also Weindling 1989.

Chapter 5. A Confusion of Terms

47 The Dicks quote is from "All for One," *New Scientist,* July 8, 2000: 30–35.

48 these little "plant-animals": Keeble's small book — *Plant Animals: A Study in Symbiosis* (Cambridge: Cambridge University Press, 1919) — is one of the most delightful I have ever read on symbiosis.

48 urged the world to take a more balanced view: Geddes and Thomson 1889.

48 it was subsequently published: R. Pound, "Symbiosis and Mutualism," *American Naturalist* 27 (1893): 509–20.

48 Pound . . . no doubt represented: *Encyclopaedia Britannica* 9: 650.

49 Pound's opinions were important: for an excellent discussion of this, see Sapp 1994.

50 the left . . . put up the most effective challenge: see Boucher 1985: 12–13; Mazumdar 1992: chap. 4.

51 "All living organisms manifest": A. Schneider, "The Phenomenon of Symbiosis," *Minnesota Botanical Studies* (1897): 923–48.

52 Merezhkovskii coined the term "symbiogenesis": Sapp 1994: 51.

52 Both Russians went on to perform a great many experiments: for a comprehensive discussion, see Sapp 1994: 49.

Chapter 6. Duels and Genes

54 "one of the most destructive episodes": Rose 1998: 38.

55 "The origin of this variation puzzled him": Mayr 1991: 46. Mayr makes this point both specifically and generally throughout this book.

56 "Why should all the parts and organs": Darwin 1859: chap. 6.

56 "You have loaded yourself with an unnecessary difficulty": Mayr 1991: 46.

56 Galton and . . . von Kölliker urged him to modify: ibid.: 46.

56 "The appeal to large variations": Ruse 1979: 206.

56 Darwin, however, refused to budge: Mayr 1979: 45–46.

57 "a scientific mistake": Agassiz quoted in Mayr 1979: 8–9.

59 "Both were convinced": R. C. Punnett, "Early Days of Genetics," *Heredity* 4 (1950): 1–10. See also Rose 1998: 39. This story may appear too perfect and quaint to avoid criticism by historians of science, but such things do happen.

61 Mendel reported his studies: G. Mendel, "Versuche über pflanzenhybriden," *Verhandlungen des naturforschenden Vereines in Brünn* 4 (1866): 3–47.

61 Nägeli's prestige . . . guaranteed: *Encyclopaedia Britannica* 14: 930; Rose 1998: 39.

61 In dismissing Mendel, Nägeli also cheated his hero: see Mayr 1979: 46; and Rose 1998: 41, 217. In perhaps the greatest irony of all, unopened copies of Mendel's papers were discovered among Darwin's effects after his death.

62 "what I say is let them fight it out": Punnett, "Early Days of Genetics."

62 even the gentlemanly art of cricket: Punnett, who recalled the lecture so vividly some forty-odd years later, prevailed upon G. H. Hardy, a mathematician at Cambridge who had "not the slightest interest in genetics" but who shared Punnett's passion for cricket, to work out the formula for heterozygous inheritance of brown and blue eyes. The formula, $pr = q^2$, is now known as the "Hardy-Weinberg equilibrium." Punnett, "Early Days of Genetics": 9–10. See also Rose 1998: 46.

62 "If we could conceive of an . . . organism": Bateson 1913: 88, quoted in Sapp 1994: 126.

63 "The voices of Haeckel, Weismann . . .": Mayr 1979: 46 and 60.

64 commonly known as . . . "neo-Darwinism": strictly speaking, the term should be "modern Darwinism" since "neo-Darwinism" was coined earlier by George J. Romanes, a later disciple of Darwin's, to accommodate Weismann's careful exclusion of any Lamarckian element in the prevailing theories around Darwinism (see Mayr 1979: 110). But after long and widespread use (Mayr might say misuse), the term has been adopted to refer to Darwinism after the synthesis. See, for example, Ridley 1997: 10.

Chapter 7. The Inner Landscape

67 "The biomarkers we report": J. J. Brocks, G. A. Logan, et al., "Archean Molecular Fossils and the Early Rise of Eukaryotes," *Science* 285 (1999): 1033–36.

67 "All living creatures . . . come about through the union": Portier 1918: vii. See also extensive discussion in Sapp 1994. For background on Portier, and for the history of symbiosis in general, throughout this book I am deeply grateful to Jan Sapp, who was kindness and courtesy itself throughout my interview and many communications.

68 Portier's book was received with incredulity: Sapp interview; Sapp 1994: chap. 6. See also Sapp, "Symbiosis and Cytoplasmic Inheritance," in Margulis and Fester 1991.

69 "Portier . . . began the controversy": Buchner 1953: 70. In this chapter, "Wrong Paths in Symbiosis Research," Buchner attacks many other early symbiologists, including Richard Altmann and Ivan Wallin.

69 "such 'errors' only served to retard": ibid.: 74.

70 In agricultural colleges such as Rutgers: the work at Rutgers is described in Ryan 1993.

71 a relationship that "seems especially equable": Thomas 1974.

71 symbiotic associations were actually driven by hostility: see Sapp 1994: 135.

71 mechanisms by which parasitism evolves: P. W. Price, "The Web of Life," in Margulis and Fester 1991.

71 hydra will survive even if the algae are bleached: Douglas 1994: 3–4.

72 common denominator . . . as a novel metabolic capability: ibid.: 9. Douglas amplified this point in an interview.

72 "one can kill aphids with antibiotics": Smith and Szathmáry 1999: 102.

73 crucial role of symbiosis: Wallin 1927.

73 Wallin . . . encountered nothing but ridicule: E. V. Cowdry and P. K. Olitsky, "Differences Between Mitochondria and Bacteria," *Journal of Experimental Medicine* 36 (1922): 521–36. The opposition to Wallin is analyzed in Sapp 1994: 117–18.

74 "mutation, reproduction and the reproduction of mutation": Muller quoted in Sapp, "Symbiosis and Cytoplasmic Inheritance," in Margulis and Fester 1991: 18. See also Sapp 1994.

Chapter 8. The Enigma of Heredity

76 "We have genes that make proteins": Sapp interview.

76 In his investigations of . . . paramecium: T. M. Sonneborn, "The Cytoplasm in Heredity," *Heredity* 4 (1950): 11–37.

78 "In a word, the cytoplasm may be ignored": Morgan quoted in Sapp 1994: 114.

79 "Time is insignificant": *Encyclopaedia Britannica* 7: 114. It is increasingly difficult to find books or articles on Lamarck. But see Jordanova 1984; and Jablonka and Lamb 1995.

81 Lederberg proposed the term "plasmid": J. Lederberg, "Cell Genetics and Hereditary Symbiosis," *Physiology Review* 32 (1952): 403–30.

81 Dobzhansky was also considering the possibility: Sapp, "Symbiosis and Cytoplasmic Inheritance": 156.

83 refuted all evidence for cytoplasmic inheritance: R. A. Raff and H. R. Mahler, "The Non-symbiotic Origin of Mitochondria," *Science* 177 (1972): 575–82.

Chapter 9. Symbiosis Comes of Age

85 "Neo-Darwinism is misleading": interview with Lynn Margulis. See also C. Mann, "Lynn Margulis: Science's Unruly Earth Mother," *Science* 252 (1991): 378–81. Mann mistakes the lecture setting as the University of Massachusetts.

85 "red tide" microorganism: see L. J. Goff, chap. 23, in Margulis and Fester 1991.

88 Mitochondria . . . have their own DNA: P. H. Raven, "A Multiple Origin for Plastids and Mitochondria," *Science* 169 (1970): 641–46; G. Attardi, "Biogenesis of Mitochondria," *Annual Review of Cell Biology* 4 (1988): 289–333; M. W. Gray and W. F. Doolittle, "Has the Endosymbiont Hypothesis Been Proven?" *Microbiology Review* 46 (1982): 1–42; D. Yang, Y. Oyaizu et al., "Mitochondrial Origins," *Proceedings of the National Academy of Sciences USA* 82 (1985): 4443–47.

89 typhus-causing germ is the closest living relative: G. E. Andersson, A. Zomorodipour, et al., "The Genome Sequence of *Rickettsia prowazeki* and the Origin of Mitochondria," *Nature* 396 (1998): 133–40; M. W. Gray, "Rickettsia, Typhus and the Mitochondrial Connection," *Nature* 396 (1998): 109–10.

90 "That's the ciliate symbiont": interview with Margulis.

92 the basis for . . . SET: L. Sagan [Margulis], "On the Origin of Mitosing Cells," *Journal of Theoretical Biology* 14 (1967): 225–74. Her first book was Margulis 1970. See also Margulis 1981.

92 theory . . . is "incomparably more inspiring": Dawkins 1995: 52–53.

93 a composite of Darwinian and symbiotic mechanisms: Smith and Szathmáry 1999: 61–66.

93 "I am prepared to be incorrect": Margulis 1998: 48. For rival theories in relation to SET, see W. F. Doolittle, "Phylogenetic Classification and the Universal Tree," *Science* 284 (1999): 2124–28; B. F. Lang, M. W. Gray, et al., "Mitochondrial Genome Evolution and the Origin of Eukaryotes," *Annual Review of Genetics* 33 (1999): 351–97; M. W. Gray, G. Burger, and B. F. Lang, "The Origin and Early Evolution of Mitochondria," *Genome Biology* 2 (2001): 1018.1–5; M. W. Gray, G. Burger, and B. F. Lang, "Mitochondrial Evolution," *Science* 283 (1999): 1476–81.

Chapter 10. The Wonder of Symbiosis

94 The Lederberg quote is from his chapter in Morse 1993: "Viruses and Humankind: Intracellular Symbiosis and Evolutionary Competition."

96 pine tree may have several different fungi: B. Kendrick, chap. 17, in Margulis and Fester 1991.

96 land plants may have arisen from . . . early incorporation: P. R. Atsatt, chap. 21, in Margulis and Fester 1991.

97 "Several of these relationships have given rise": Kendrick, chap. 17, in Margulis and Fester 1991.

97 characteristics that define each phylum: D. Bermudes and R. C. Back, "Symbiosis Inferred from the Fossil Record," in Margulis and Fester 1991. See also D. Bermudes and L. Margulis, "Symbiont Acquisition as Neoseme: Origin of Species and Higher Taxa," *Symbiosis* 4 (1987): 185–98.

98 described the hydroid *Myrionemia amboinense:* R. Law, "The Symbiotic Phenotype: Origins and Evolution," in Margulis and Fester 1991.

98 simplicity appeals more to our human aesthetic: McMullin in Ruse 1999: 33.

Chapter 11. Planetary Evolution

99 the concept of "paradigm": Kuhn 1962.

100 "they are often intolerant": ibid.: 24.

101 "I was even more fascinated": interview with James Lovelock, Aug. 2001.

106 Even Carl Sagan disagreed with him: ibid.

106 offering to publish his lecture: J. R. Lovelock and L. Margulis, "Atmospheric Homeostasis by and for the Biosphere: The Gaia Hypothesis," *Tellus* 16 (1974): 2–10.

106 Holland . . . exclaimed, "We don't need Gaia": Lovelock interview. See also Lovelock 1988: 31–32.

107 Gaia as an "evil religion": ibid.

108 "a synthetic balance-of-nature theory": F. N. Egerton, "Changing Concepts of the Balance of Nature," *Quarterly Review of Biology* 48 (1973): 322–50.

108 "About thirty years ago there was much talk": Darwin quoted in Shermer

2001: 119. Shermer also gives an abbreviated version in his article in *Scientific American* (Apr. 2001): 26.

108 redefining "Biosphere": Vernadsky 1997.

110 Vernadsky . . . estimated the organic carbon content: ibid.: 106.

111 Gaia as "the series of interacting ecosystems": this perspective was first published in Lovelock and Margulis, "Atmospheric Homeostasis."

111 "It gave me a feeling there were tramlines": Lovelock quoted in M. Bond, "Father Earth," *New Scientist* (Sept. 9, 2000): 44–47.

112 Gaia was dismissed in contemptuous terms: Lovelock 2001: 250.

112 "If Daisyworld is valid": ibid.: 250.

112 Gould, Smith, Mayr criticize Gaia: as quoted in Mann, "Lynn Margulis."

Chapter 12. Redefining Concepts

115 The Villarreal quote is from my interview with him.

116 "The crucial step . . . is to understand the mechanism": Smith and Szathmáry 1999: 1.

117 "soggy semantics": D. Lewis, "Symbiosis and Mutualism: Crisp Concepts and Soggy Semantics," in Boucher 1985.

117 Margulis subdivides . . . symbiosis: Margulis 1981: chap. 7.

117 "interaction of dissimilar genomes": W. Reisser, "Symbiosis Redefined: Symbiotic Features of Virus-Host Interactions," *Symbiosis* 14 (1992): 83–86. I have also communicated personally with Professor Reisser on this topic.

Chapter 13. That First Spark

123 Erasmus Darwin (1731–1802), the grandfather of both Charles Darwin and Francis Galton, put forward in *Zoonomia or the Laws of Organic Life* (1794–96) evolutionary views that were essentially Lamarckian. His overly simple conclusions, based on his own observations of life, were rejected by Charles.

123 It was nevertheless remarkable for what it did reveal. See entire issues of *Nature* and *Science* for the week of Feb. 17, 2001.

123 it's not how many genes you've got: see *New Scientist* (Feb. 17, 2001): 6.

124 life forms from space are constantly arriving: Hoyle and Wickramasinge 1993.

125 Meteorites . . . bear . . . organic compounds: M. Bergstein, S. A. Sandford, and L. J. Allamandola, "Life's Far-Flung Raw Materials," *Scientific American* (July 1999): 26–33.

126 "If . . . in some warm little pond": Darwin quoted in Fortey 1997: 37.

126 a more complex explanation involving information theory: P. Davies, "Life Force," *New Scientist* (Sept. 18, 1999): 27–30.

127 discovery of the first RNA-based enzymes: Smith and Szathmáry 1999: 39; J. Horgan, "The World According to RNA," *Scientific American* (Jan. 1996): 16–17. For detailed background, see Dyson 1999.

128 there must have been an "RNA world": W. Gilbert, "The RNA World," *Nature* 319 (1986): 618. See also Dyson 1999: 11–14.

128 Harold Urey: Urey would later receive a Nobel Prize for the discovery of deuterium.

128 contents were found to include amino acids: Davies 1998: 56–63. See also *Encyclopaedia Britannica* 22: 993.

130 some elegant mathematical extrapolations: K. H. Keeler, "Cost Benefit Models of Mutualism," in Boucher 1985.

130 "The origin of the code assumed co-operative interactions": Smith and Szathmáry 1999: 47.

131 "double origin" of life: Dyson 1999.

132 evolution of life "is a cosmic imperative": de Duve quoted in Davies, "Life Force." See also de Duve 1996.

Chapter 14. The True Pioneers

133 Reeve quoted in Highfield, "Scientists Find Life but Not as We Know It," *Daily Telegraph* (Aug. 24, 1996).

134 The bacterium *Haloarcula marismortui:* H. Watzman, "The Secret of Life in the Dead Sea," *New Scientist* (June 22, 1996): 15.

134 *Thiobacillus thiooxidans* can digest concrete: B. Crystall, "Bug custard: Concrete-chomping Bacteria That Could Clean Up Nuclear Spills," *New Scientist* (Oct. 9, 1999): 6.

136 "Our hypothesis is that life is a cosmic phenomenon": Hoyle and Wickramasinge 1993. See also M. Chown, "Seeds, Soup and the Meaning of Life," *New Scientist* (Aug. 17, 1996): 6; J. Hecht, "Life Will Find a Way," *New Scientist* (Mar. 17, 2001): 4; J. Marchant, "Life from the Skies," *New Scientist* (July 15, 2000): 4. See also Bergstein et al., "Life's Far-Flung Raw Materials."

136 hitherto unknown species of bacterium: A. Coghlan, "Sleeping Beauty," *New Scientist* (Oct. 21, 2000): 12. See also J. Knight, "The Immortals," *New Scientist* (Apr. 28, 2001): 36–39.

136 even older rocks . . . contain the fingerprints: S. J. G. Mojzsis, K. D. Arrhenius, et al., "Evidence for Life on Earth Before 3800 Million Years Ago," *Nature* 384 (1996): 55–59.

137 Woese . . . pioneered new genetic techniques: C. R. Woese, "Archaebacteria," *Scientific American* 244 (1981): 94–106. See also B. Holmes, "Life Unlimited," *New Scientist* (Feb. 10, 1996): 26–29.

138 a radical shakeup of the taxonomic divisions: C. R. Woese, O. Kandler, and M. L. Wheelis, "Towards a Natural System of Organisms: Proposal for the Domains Archaea, Bacteria, and Eukarya," *Proceedings of the National Academy of Sciences USA* 87 (1990): 4576–79.

140 a single bacterium . . . would . . . produce 2^{288} copies: Margulis 1997: 75.

141 a very brief paper: J. Lederberg, E. L. Tatum, "Gene Recombination in *Escherichia coli,*" *Nature* 158 (1946): 558.

142 This transfer of resistance is mediated by a . . . plasmid: A. Guiyoule, G. Gerbaour, et al., "Transferable Plasmid-mediated Resistance to Streptomycin in a Clinical Isolate of *Yersinia pestis,*" *Emerging Infectious Diseases* 7 (2001): 43–48.

142 all the strains of bacteria . . . form a single pool: S. Sonea, "Bacterial Evolution without Speciation," in Margulis and Fester 1991: 95–105.

143 This . . . "is a perfect example of the Gaia hypothesis": ibid. "In an evolutionary innovation unprecedented": Margulis 1997: 101.

Chapter 15. The Greening of the Earth

147 the trials and errors that proliferated: Raven, "A Multiple Origin."

148 invasion of dry land by the kingdom of plants: McMenamin and McMenamin 1994: 4.

148 chloroplast genome of . . . *Mesostigma viride:* C. Lemieux, C. Otis, and M. Turmel, "Ancestral Chloroplast Genome in *Mesostigma viride* Reveals an Early Branch of the Green Plant Evolution," *Nature* 403 (2000): 649–52.

149 perhaps the greatest need . . . was water itself: McMenamin and McMenamin 1994.

150 when the going gets tough: ibid.: 67–73.

151 "Those first plants began in water": my interview with David Read.

154 mycorrhizae have been called "the universal symbiosis": ibid.; see also T. H. Nicholson, "Vesicular-arbuscular Mycorrhiza — a Universal Plant Symbiosis," *Science Progress* 55 (1967): 561–81.

154 the evolution of . . . broad leaves: D. J. Beerling, C. P. Osborne, and W. G. Chaloner, "Evolution of Leaf-Form in Land Plants Linked to Atmospheric CO_2 Decline in the Late Paleozoic Era," *Nature* 410 (2001): 352–54.

155 the evolution of the embryo: P. R. Atsatt, chap. 21, in Margulis and Fester 1991: 302.

156 the biomass of life on land: McMenamin and McMenamin 1994: 22.

157 flowers and fruits may be the results of symbioses: K. A. Pirozynski, chap. 24, in Margulis and Fester 1991.

157 the first plant to have its genome mapped: the Arabidopsis Genome Initiative, "Analysis of the Genome Sequence of the Flowering Plant *Arabidopsis thaliana,*" *Nature* 408 (2000): 796–813; G. Vines, "Wonderweed," *New Scientist* (Dec. 2, 2000): 36–39; A. Coghlan, "It's a Wonderful Weed," *New Scientist* (Dec. 16, 2000): 14–15; R. Highfield, "This Little Weed Is a Genetic Giant," *Daily Telegraph* (Dec. 14, 2000): 24.

Chapter 16. The Creatures of Land, Air, and Sea

160 Bioluminescence had been observed . . . in fish: M. J. McFall-Ngai, chap. 25, in Margulis and Fester 1991.

161 "We discovered extraordinary communities": J. B. Corliss, J. Dymond, et al., "Submarine Thermal Springs on the Galapagos Rift," *Science* 203 (1979): 1073–83; R. D. Vetter, "Symbiosis and the Evolution of Novel Trophic Strategies: Thiotrophic Organisms at Hydrothermal Vents," in Margulis and Fester 1991.

162 in the depths of a cave in Rumania: "The Secret Underworld," 1997 television documentary on the discovery of the Movile cave in Rumania.

162 bacterial life existed long before . . . the Cambrian: McMenamin and McMenamin 1994: 47–49. See also McMenamin 1998.

163 "Self-sufficient feeding strategies meant . . .": McMenamin and McMenamin 1994: 47–49.

163 life as evolving in five major transitions: Smith and Szathmáry 1999: 107.

164 different sets of key genes are active: Monod 1997.

166 Recent research on *Wolbachia:* Sapp, "The Evolution of Complexity," *History and Philosophy of the Life Sciences* 21 (1999): 215–26; J. H. Warren, "Biology of *Wolbachia,*" *Annual Review of Entomology* 42 (1997): 587–609; T. Wenseelers et al., "Widespread Occurrence of the Micro-organism *Wolbachia* in Ants," *Proceedings of the Royal Society of London B* 265 (1998): 1447–52.

167 "So important are insects": Wilson 1992: 125.

168 early Cambrian fossils in . . . the Burgess shale: Gould 1989: 154.

169 coelacanth, which had been assumed extinct: Weinberg 1999.

169 importance of endosymbiotic cellular evolution: Raven, "A Multiple Origin."

Chapter 17. A New Tree of Evolution

171 a fourth kingdom, the Protoctista: Copeland 1956.

172 fungi . . . warranted a kingdom of their own: R. H.Whittaker, "On the Broad Classification of Organisms," *Quarterly Review of Biology* 34 (1959): 210–26.

172 "a twisted, tangled, pulsing entity": Margulis and Schwartz 1982.

173 a three-"domain" classification: Woese, Kandler, and Wheelis, "Towards a Natural System of Organisms."

Chapter 18. The Cycles of Life

175 Earth froze into a gigantic snowball: P. F. Hoffman and D. Shrag, "Snowball Earth," *Scientific American* (Jan. 2000). See also R. Highfield, in *Daily Telegraph* (Dec. 19, 2001): 14.

177 "the accumulation of small change": Dawkins 1986.

178 "the next breath you inhale": P. Cloud and G. Aharon, "The Oxygen Cycle," *Scientific American* 223 (Sept. 1970): 110–23.

180 an advantageous mutation enabled grasses: Lovelock 1988: 128–30.

180 the sole source of carbon dioxide was volcanic emissions: Holland 1984.

181 This contribution of Gaian theory to the carbon dioxide cycle: J. E. Lovelock and A. J. Watson, "The Regulation of Carbon Dioxide and Climate: Gaia or Geochemistry?" *Journal of Planetary and Space Science* 30 (1982): 795–802.

182 dimethyl sulfide was present: J. E. Lovelock, R. J. Mags, and R. W. Rasmussen, "Atmospheric Dimethyl Sulphide and the Natural Sulphur Cycle," *Nature* 237 (1972): 452–53.

182 sulfur cycle is carried out through dimethyl sulfide: P. S. Liss and P. G. Slater, "Flux of Gas across the Air-Sea Interface," *Nature* 247 (1974): 181–84.

184 Dimethyl sulfide . . . provided the condensation nuclei: R. J. Charlson, J. E. Lovelock, M. O. Andreae, and S. J. Warren, "Oceanic Phytoplankton, Atmospheric Sulphur, Cloud Albedo and Climate," *Nature* 326 (1987): 655–61.

186 "You have convinced me that the Earth is self-regulating": Lovelock interview.

186 the importance . . . of finding ways to spread their seeds: W. D. Hamilton and T. M. Lenton, "Spora and Gaia," *Reviews of Geophysics* 27, no. 2 (1998): 215–22.

186 "We wait for an Isaac Newton": Lovelock interview.

186 "The Earth System behaves as a single self-regulating system": Declaration of the Geophysical Union, Amsterdam, July 2001. See "Challenges of a Changing Earth" at www.sciconf.igbp.kva_se/p1.html.

Chapter 19. The Sound and the Fury

188 Creationists . . . "with their usual skill . . .": Gould described his perspective briefly in "Opus 200," *Natural History* 8 (1991): 12–18. See also Gould 1983. Gilkey 1985 provides a theological perspective.

189 colleagues are "always circling": Ruse 1999.

189 Margulis derides neo-Darwinism: Mann, "Lynn Margulis," chap. 9.

189 "Symbiosis is not an alternative": Smith and Szathmáry 1999: 107.

189 "Why then is not every geological formation . . . ?": Darwin 1859: chap. 6.

190 when a small group . . . became geographically isolated: Mayr 1963, 1991.

190 sudden changes . . . in fossils of Paleozoic invertebrates: N. Eldredge, "The Allopatric Model and Phylogeny in Paleozoic Invertebrates," *Evolution* 25 (1971): 156–67.

190 Gould was assigned to talk about speciation: Gould, "Opus 200."

190 lecture . . . later became a chapter: N. Eldredge and S. J. Gould, "Punctuated Equilibrium: An Alternative to Phyletic Gradualism," in *Models in Palaeobiology*, ed. T. J. M. Schopf (San Francisco: W. H. Freeman, 1972).

191 Gould . . . showed the article to his father: Gould, "Opus 200."

191 Dawkins dismissed it as unnecessary: Dawkins 1986, chap. 9.

191 Mayr made essentially the same point: Mayr 1991: 153.

191 "Most of my testimony . . . centered upon the creationists' distortion": Gould, "Opus 200."

192 Judge Overton's decision . . . was upheld: Gould 1991: 431. See also Gould 1983: chap. 21.

192 "We believe that punctuational change dominates": S. J. Gould and N. Eldredge, "Punctuated Equilibria: The Tempo and Mode of Evolution Reconsidered," *Palaeobiology* 3 (1977): 125–51.

192 endosymbiosis . . . is "one possible mechanism": D. Bermudes and R. C. Back, "Symbiosis Inferred from the Fossil Record," in Margulis and Fester 1991.

192 symbiosis may have given rise to three of the five: Smith and Szathmáry 1999: 107.

193 Eldredge could think of only a single example: Margulis 1998: 8. In conversation, Margulis confirmed that the parasite was a mycoplasma.

193 "symbiosis certainly challenges individualism": Sapp interview, 2001.

193 studies . . . could be usefully reappraised: D. H. Boucher, "The Idea of Mutualism, Past and Future," in Boucher 1985.

194 "ecologists had neglected to look for mutualism": quoted ibid.: 6.

194 endosymbiosis gave rise to at least twenty-eight . . . phyla: D. Bermudes and L. Margulis, "Symbiont Acquisition as Neoseme: Origin of Species and Higher Taxa," *Symbiosis* 4 (1987): 185–98.

194 symbiosis might have contributed to . . . marine fossil groups: Bermudes and Bach, "Symbiosis Inferred."

194 allocated them to species on the basis of their symbiotic partners: Margulis interview.

195 all embryos and larvae originated by hybridization: D. I. Williamson, "Larval Transfer and the Origins of Larvae," *Zoological Journal of the Linnaean Society* 131 (2001): 111–22; D. I. Williamson and A. L. Rice, "Larval Evolution in the Crustacea," *Crustaceana* 69, no. 3 (1996): 267–87.

Chapter 20. Sex and Cannibalism

198 an elegant and mesmeric dance: For more information on this dance and on hummingbirds, see A. L. Austin et al., *Birds of the World* (London: Hamlyn, 1962). See also R. Attenborough, *The Trials of Life* (London: Collins/BBC, 1990).

203 meiotic sex . . . as cannibalism: Margulis told me, "Cleveland would be appalled at the neglect of his basic life's work, which was to discover the origin of chromosomes, mitosis, and sexuality." Cleveland's most important papers are "Symbiosis among Animals with Special Reference to Termites and Their Intestinal Flagellates," *Quarterly Review of Biology* 1 (1926): 51–56; and "The Origin and Evolution of Meiosis," *Science* 105 (1947): 287–89.

203 Cleveland entertained his audience with films: One of Cleveland's films, "Sexuality and Other Features of the Flagellates of *Cryptocercus,*" is now owned by the Department of Zoology, University of Massachusetts, Amherst.

204 "those that couldn't encyst turned to cannibalism": Margulis interview. She has written a four-page paper on the life of Cleveland, "L. R. Cleveland: Scientist Misunderstood," but has not been able to get it published.

204 the species as a whole is an evolving unit: Smith and Szathmáry 1999: 79.

205 a Darwinian scenario for the evolution of meiotic sex: ibid.: 87–90.

206 mutualistic symbiosis tends to protect the partnership: R. Law, "Evolution in a Mutualistic Environment," in Boucher 1985.

206 human mitochondrial genome is the smallest: G. Attardi and G. Schatz, "Biogenesis of Mitochondria," *Annual Review of Cell Biology:* 4: 289–333.

208 mitochondria are sensitive to certain antibiotics: B. Storrie and G. Attardi, "Modes of Mitochondrial Formation in Hela Cells," *Journal of Cell Biology* 56: 833–38.

208 researchers attributed this loss to damage affecting the mitochondria: A. Motluk, "Foetal Exposure: X-rays in the Womb May Lead to Mental Illness in Adulthood," *New Scientist* 14 (Nov. 18, 2000).

208 culling may be a programmed elimination of eggs: P. Cohen, "The Force," *New Scientist* (Feb. 26, 2000): 30–35.

208 reducing mutational load of mitochondrial genes: L. Hurst, "Selfish Genetic Elements and Their Role in Evolution: The Evolution of Sex and Some of What That Entails," *Philosophical Transactions of the Royal Society of London B* 349 (1995): 321–32.

208 mitochondria in mammalian sperm are actively destroyed: P. Sutovsky et al., "Ubiquitin Tag for Sperm Mitochondria," *Nature* 402 (1999): 371–72.

209 Mutations in mitochondrial genes may also play a critical part: P. A.

Kiberstis, "Mitochondria Make a Comeback," *Science* 283 (1999): 1475; Gray, Burger, and Lang, "Mitochondrial Evolution." See also Cohen, "The Force."

209 "mitochondria . . . don't function efficiently in zero gravity": quoted in C. Seife, "Space Healing," *New Scientist* (Sept. 29, 1999): 20.

Chapter 21. Playing Hardball with Evolution

210 "We all play hideous games": Hopkins made this comment in an interview in the *Sunday Times* (London) *Colour Supplement* (Apr. 12, 1998).

210 "Lederberg's discoveries made bacteria as important": *Encyclopaedia Britannica* 7: 233.

211 introduced the word "plasmid": J. Lederberg, "Cell Genetics and Hereditary Symbiosis," *Physiology Review* 21 (1952): 133–39.

211 "The notion of infection was antithetical": Lederberg interview.

211 the symbiotic interaction is dynamic: R. Dubos and A. Kessler, "Integrative and Disintegrative Factors in Symbiotic Associations," in Nutman and Moss 1963.

212 the placental, or eutherian, mammals: see J. D. Archibald, A. O. Averianov, and E. G. Ekdale, "Late Cretaceous Relatives of Rabbits, Rodents, and Other Extant Eutherian Mammals," *Nature* 414 (2001): 62–65.

213 viruses in . . . perfectly healthy baboons: D. Fox, "Why We Don't Lay Eggs," *New Scientist* (June 12, 1999): 26–31. P. Cohen, "Mother's Little Helper: Viruses May Cause Disease, but Would We Be Here Without Them?"

213 Similar viruses . . . in the placentas of cats, mice: J. R. Harris, "Placental Endogenous Retrovirus (ERV): Structural, Functional, and Evolutionary Significance," *Bioessays* 20 (1998): 307–16.

214 retroviruses . . . may . . . protect the fetus: Fox, "Why We Don't Lay Eggs."

214 the placenta no longer implanted: ibid. Also my interview with Villarreal.

215 has found in the placenta a retrovirus: Fox, "Why We Don't Lay Eggs."

215 "Could an ancient endogenous retrovirus . . . ?": Harris, "Placental Endogenous Retrovirus."

215 research . . . identified one of the viral genes: S. Mi, X. Lee, et al., "Syncytin Is a Captive Retroviral Envelope Protein Involved in Human Placental Morphogenesis," *Nature* 403 (2000): 785–89.

215 evidence that preeclampsia is accompanied by a dramatic reduction: X. Lee, J. C. Keith Jr., et al., "Downregulation of Placental Syncytin Expression and Abnormal Protein Localization in Pre-eclampsia," *Placenta* 22, no. 10 (2001): 808–12.

215 preeclampsia was associated with low syncytin levels: I. Knerr, E. Beinder, and W. Rascher, "Syncytin, a Novel Human Endogenous Retroviral Gene in Human Placenta: Evidence for Its Dysregulation in Pre-eclampsia and HELLP Syndrome," *American Journal of Obstetric Gynecology* 186 (2002): 210–13.

217 "Persisting viruses can have major evolutionary consequences": Professor Villarreal kindly sent me a copy of his talk, "Role of Persisting DNA Viruses and Retroviruses in Host Evolution."

217 retroviral genes . . . linked to genetic susceptibility: B. Furlow, "The Enemy Within," *New Scientist* (Aug. 19, 2000): 38–41. See also Harris, "Placental Endoge-

nous Retrovirus"; R. Löwer, J. Löwer, and R. Kurth, "The Viruses in All of Us: Characteristics and Biological Significance of Human Endogenous Retrovirus Sequences," *Proceedings of the National Academy of Sciences USA* 93 (1996): 5177–84.

217 roles played by viruses in humans: Löwer, Löwer, and Kurth, "The Viruses in All of Us."

218 DNA polymerases make possible: W. H. Elliott and D. C. Elliott 2001.

218 B-polymerases in eukaryotes: L. P. Villarreal and V. R. DeFilippis, "A Hypothesis for DNA Viruses as the Origin of Eukaryotic Replication Proteins," *Journal of Virology* 74 (2000): 7079–84.

219 the eukaryotic nucleus began as an infection: P. J. Bell, "Viral Eukaryogenesis: Was the Ancestor of the Nucleus a Complex DNA Virus?" *Journal of Molecular Evolution* 53 (2001): 251–56. See also R. Nowak, "One Small Step: Did an Invading Virus Kick-start Our Evolution?" *New Scientist* (Sept. 15, 2001): 16.

220 compelling case . . . that viruses have played a role in evolution: Professor Villarreal kindly let me read Chapter 6 of the book prior to publication.

221 I have coined the term "aggressive symbiosis": Ryan 1997.

221 Aggressive symbioses are not confined to virus-host relationships: for grasses, see P. W. Price, "The Web of Life: Development over 3.8 Billion Years of Trophic Relationships," in Margulis and Fester 1991. For bacteria and worms, see K. H. Nealson, chap. 15, in Margulis and Fester 1991.

222 Ichneumon wasps . . . provide an even more startling example: K. Edson and D. B. Stolz, "Virus in a Parasitoid Wasp: Suppression of the Cellular Immune Responses in the Parasitoid's Host," *Science* 211 (1980): 582–83.

222 if he knew the genome of a virus that coevolved: my interview with Professor Yates for *Virus X*.

224 schizophrenia . . . may be associated with the most intelligent: Horrobin 2001. See also R. McKie, "Schizophrenia Helped the Ascent of Man," *Observer* (Mar. 18, 2001): 6.

224 retrovirus in . . . people who developed . . . schizophrenia: R. Yolken, H. Karlsson, et al., *Proceedings of the National Academy of Sciences USA* 8 (2001): 4634–39.

225 "the most omnipresent . . . interaction": D. H. Janzen, "The Natural History of Mutualisms," in Boucher 1985.

225 Endosymbiotic interactions have given rise: Douglas Fox, "Wallaby Nations," *New Scientist* (Aug. 3, 2002): 32–35. *See also* R. J. W. O'Neill, M. J. O'Neill, and J. A. M. Graves, "Undermethylation Associated with Retroelement Activation and Chromosome Remodelling in an Interspecific Mammalian Hybrid," *Nature* 393 (1998): 68–72.

Chapter 22. People: The Most Mutualistic of Organisms?

226 *Australopithecus garhi:* B. Asfaw, T. White, et al., "*Australopithecus garhi:* A New Species of Early Hominid from Ethiopia," *Science* 284 (1999): 629–35.

228 "organic substances which the body requires": Passmore and Eastwood 1986: 132.

229 our dependency on . . . omega-3 fats: see F. Ryan and R. Saynor, *The Eskimo Diet* (London: Ebury Press, 1990).

229 vitamin D . . . may lower the risk of juvenile-onset diabetes: E. Hyppönen, E. Läärä, et al., "Intake of Vitamin D and the Risk of Type 1 Diabetes: A Birth-Cohort Study," *Lancet* 358 (2001): 1500–3. See also editorial in the same issue: 1476–77.

230 "The observed decline" [in spina bifida]: Honein quoted in A. Coghlan, "Fortified Food Staves Off Spina Bifida," *New Scientist* (June 30, 2001): 17.

230 "animals . . . as frills around a compost heap": D. H. Janzen, "The Natural History of Mutualisms," in Boucher 1985.

231 our gut flora . . . "may even be able to do some things": ibid.

232 "The processes of digestion . . . are linked": Cunningham-Rundles quoted in D. Derbyshire, "Bugs at War in the Body," *Daily Telegraph* (July 26, 2000): 25.

232 parasitic worms . . . alter the pattern: R. McKie, "Now the Doctors Say Parasitic Worms Are Good for You," *Observer* (May 13, 2000): 8.

232 "live" yogurt . . . can help . . . severe diarrhea: "Effect of Long-term Consumption of Probiotics Milk on Infections in Children Attending Day Care Centers: Double Blind, Randomized Trial," *British Medical Journal* 322 (2001): 1327–29; D. James, "Do Probiotics Prevent Childhood Illnesses?" *British Medical Journal* 322 (2001): 1318–19.

233 Evolutionists . . . are looking beyond natural selection: E. Sober and D. S. Wilson, *Unto Others* (Cambridge, Mass.: Harvard University Press, 1999); W. Swenson, D. S. Wilson, and R. Elias, *Proceedings of the National Association of Sciences* 97 (2000): 9110–14. *See also L. Dicks, New Scientist* (July 2000): 30–35.

233 "Together with its symbionts/parasites": Joshua Lederberg, personal communication.

234 how we . . . need to change the way we think: Janzen has been working for many years on conservation in the 153,000 hectares of tropical dry forest known as the Area de Conservación Guanacaste, a World Heritage Site in northwestern Costa Rica. He has adopted an unusual approach to global conservation in assuming that people will care more if they regard all remaining wilderness areas as "their own gardens." See D. H. Janzen, "Good Fences Make Good Neighbors," *PARKS* 11, no. 2 (2001): 41–49; D. H. Janzen, "Lumpy Integration of Tropical Wild Biodiversity with Its Society," in *A New Century of Biology,* ed. W. J. Kress and G. W. Barrett (Washington: Smithsonian Institution Press, 2001): 133–48.

234 "The largest single step in the ascent of man": J. Bronowski, *The Ascent of Man* (Boston: Little, Brown, 1976).

236 differences in brain power . . . reflect . . . size of the brain: M. A. Crawford, S. C. Cunnane, and L. S. Harbige, "A New Theory of Evolution: Quantum Theory," in *Third International Congress on Essential Fatty Acids and Eicosanoids,* eds. A. J. Sinclair and R. Gibson (Adelaide, Australia: AOCS Press, 1993): 87–95.

236 relative brain size actually decreased: Crawford, Cunnane, and Harbige, "A New Theory"; R. D. Martin, "Human Brain Evolution in an Ecological Context," Fifty-Second James Arthur Lecture on the Evolution of the Human Brain (New York: American Museum of Natural History, 1983).

237 australopithecines showed little increase in brain size: M. A. Crawford, M.

Bloom, et al., "Evidence for the Unique Function of Docosahexanoic Acid During the Evolution of the Modern Hominid Brain," *Lipids* 34 (1999): 539–47; H. M. Su, L. Bernardo, et al., *Lipids* 34 (1999): 5347–50; G. C. Conroy, G. W. Weber, et al., "Endocranial Capacity in an Early Hominid Cranium from Sterkfontein, South Africa," *Science* 280 (1998): 1730–31.

237 the researchers do not believe that *Homo sapiens:* Crawford, Cunnane, and Harbige, "A New Theory"; N. E. Sikes, "Early Hominid Habitat Preferences in East Africa: Paleosol Carbon Isotopic Evidence," *Journal of Human Evolution* 27 (1994): 25–45; C. L. Broadhurst, S. C. Cunnane, and M. A. Crawford, "Rift Valley Lake Fish and Shellfish Provided Brain-specific Nutrition for Early *Homo,*" *British Journal of Nutrition* 79 (1998): 3–21.

237 so large "as to imply": Crawford and Bloom, "Evidence for the Unique Function."

238 preterm delivery and low birth weights were reduced: S. F. Olsen and N. J. Secher, "Low Consumption of Seafood in Early Pregnancy as a Risk Factor for Preterm Delivery: Prospective Cohort Study," *British Medical Journal* 324 (2002): 447–50.

Chapter 23. From Cuddling Fish to Bartering Cities

240 outcome . . . will be determined by cultural processes: E. O. Wilson 1978: 103–5.

240 "the ideal of the ethical man . . .": T. H. Huxley, "The Struggle for Existence: A Programme," *Nineteenth Century* 23, no. 32 (1888): 161–80.

242 Lynn Dicks challenge "If you accept that evolution is all about selfish genes": L. Dicks, "All for One," *New Scientist* (July 8, 2000): 30–35.

242 The biologists were interested in cooperative behavior: R. Bshary and M. Würth, "Cleaner Fish *Labroides dimidiatus* Manipulate Client Reef Fish by Providing Tactile Stimulation," *Proceedings of the Royal Society of London* 268 (2001): 1495–1501.

243 clients visit cleaners to be stroked: G. S. Losey, "Fish Cleaning Symbiosis: Proximate Causes of Host Behaviour," *Animal Behaviour* 27 (1979): 669–85; G. S. Losey, "Cleaning Symbiosis," *Symbiosis* 4 (1987): 229–58.

243 "the systematic study of the biological basis of behavior": Wilson 1975.

243 "human emotional responses . . . have been programmed": Wilson 1978: 6.

244 culture itself was ultimately inherited: ibid.: 21.

244 differences . . . that "cannot reasonably be explained": ibid.: 48–49.

244 Wilson's proposals . . . provoked an outraged reaction: Sapp interview. The controversy was brought together in *The Sociobiology Debate,* edited by Arthur Caplan. See also Ruse 1998: chap. 9; R. C. Lewontin, "Sociobiology — A Caricature of Darwinism," in *PSA 1976,* ed. F. Suppe and P. Asquith (East Lansing, Mich.: Philosophy of Science Association, 1977).

245 complex social interaction was the reason for the evolution: R. A. Barton and R. I. M. Dunbar, "Evolution of the Social Brain," in *Machiavellian Intelligence II,* ed. A. Whiten and R. W. Byrne (Cambridge: Cambridge University Press, 1997).

245 cooperation is the equivalent between members of the same species:

D. Lewis, "Symbiosis and Mutualism: Crisp Concepts and Soggy Semantics," in Boucher 1985.

247 Darwin's followers completely ignored the role of cooperation: Singer 1999: 19. J. M. Smith, foreword to H. Cronin, *The Ant and the Peacock* (Cambridge: Cambridge University Press, 1991): xx. See also Dugatkin 1999: 28; Darwin 1872.

247 "kinship theory": Dugatkin 1999: 40.

247 the mathematics of risk versus gain: W. D. Hamilton, "The Genetical Evolution of Social Behaviour," parts I and II, *Journal of Theoretical Biology* 7 (1964): 1–16; 17–52.

247 cooperation between unrelated individuals: R. L. Trivers, "The Evolution of Reciprocal Altruism," *Quarterly Review of Biology* 46 (1971): 189–226.

248 Trivers . . . lacked mathematical proof: Dugatkin 1999: 83.

249 altruism . . . "an ignoble, low substitute": Dawkins, foreword to Axelrod 1984.

250 Axelrod and Hamilton published their findings: R. Axelrod and W. D. Hamilton, "The Evolution of Cooperation," *Science* 212 (1981): 1390–96; Axelrod 1984.

250 "Once the genes for cooperation exist": Axelrod 1984: 97.

250 the emergence of modern human society required cooperation: Smith and Szathmáry 1999: 148.

250 "Each one of us is a community": Dawkins 1995: 52.

251 The core of a mutualism is the payment: D. H. Janzen, "The Natural History of Mutualisms," in Boucher 1985: 58–59.

251 "jumping genes," which replicate out of phase: Smith and Szathmáry 1999: 97.

251 selection also acts at the level . . . of the organism as a whole: Mayr 1991: 142–43. See also Smith and Szathmáry 1999: 95.

252 selection at the group level . . . might explain mutual support: Darwin 1871: chap. 5. See also Sober and Wilson 1998: 4ff.

252 a similar group-selection argument: Smith and Szathmáry 1999: chap. 9. See also J. M. Smith, "A Darwinian View of Symbiosis," in Margulis and Fester 1991.

253 Rose has accused the "sexual Darwinists" of having a "Flintstones' view": Rose is quoted in D. Hill, "Back to the Stone Age," *Observer* (Feb. 27, 2000).

254 "rape . . . should be viewed as a natural biological phenomenon": D. Concar, "Crimes of Passion: An Interview with Randy Thornhill," *New Scientist* (Feb. 19, 2000): 45–47. See also R. Thornhill and C. Palmer *A Natural History of Rape* (Cambridge, Mass.: MIT Press, 2000).

254 "Why do people devote so much time . . . ?": Buss 1994.

254 cooperative behavior "in everything from worms to humans": W. C. Allee, "Where Angels Fear to Tread: A Contribution from General Sociology to Human Ethics," *Science* 97 (1943): 517–25. See also Dugatkin 1999: 140–41.

255 chorusing behavior of songbirds was a means of censuring: Wynne-Edwards 1962.

255 detailed genetic models of . . . group selection: see Dicks, "All for One."

255 Group selection may explain the evolution of . . . honeybees: T. Seeley, *The*

Wisdom of the Hive (Cambridge, Mass.: Harvard University Press, 1995). See also Dugatkin 1999: 149–50.

255 The relevance to human society . . . becomes obvious: Smith and Szathmáry 1999: 145ff.

255 Hutterites . . . work together: Dugatkin 1999: 158–63.

256 "A relatively bizarre form of locomotion": Lovejoy quoted in P. Martin, "The Bone People," *Sunday Times* (London) color supplement, Apr. 23, 1999. See also Asfaw, White, et al., *"Australopithecus garhi."*

258 egalitarian behavior leading to deep-seated cooperation: Dickens 2000: 87.

258 "An evolved collaboration in resource use": ibid.: 88.

258 The pyramids were found to mark the site of Caral: my communication with Dr. Ruth Shady of San Marcos University. See also the BBC *Horizon* program "The Lost Pyramids of Caral." The Web site www.bbc.co.uk/science/horizon/2001/caral.shtml has links to the National University in Peru and also to the archaeological museum at the University of Minnesota.

Chapter 24. The Weave of Life

261 The Dubos quote is from his article "Symbiosis Between the Earth and Humankind," *Science* 193 (1976): 459–62.

262 credit goes to the man . . . who convinces the world: Pasteur quoted in R. Dubos and J. Dubos, *The White Plague* (London: Victor Gollancz, 1953).

263 when it comes to symbiosis, natural selection has shortcomings: R. Law, "Evolution in a Mutualistic Environment," in Boucher 1985.

264 selection probably operates at the level of the partnership: J. M. Smith, "A Darwinian View of Symbiosis," in Margulis and Fester 1991.

264 One possible mechanism for survival: K. J. Jeon, "Symbiosis of Bacteria with Amoeba," in *Cellular Interactions in Symbiosis and Parasitism,* ed. C. B. Cook, P. W. Pappas, and E. D. Rudolph (Columbus: Ohio State University Press, 1980): chap. 11. See also Margulis and Schwartz 1998: 118–21.

265 Darwinism is "heterogeneous": Mayr 1991: 143.

266 "a highly complex research program": ibid.

266 "The combination . . . makes further progress . . . possible": W. Schwemmler, "Symbiogenesis in Insects as Models for Cell Differentiation, Morphogenesis, and Speciation," in Margulis and Fester 1991.

266 "No serious scholar . . . thinks that race is a scientific concept": Venter quoted in R. Highfield, "Science Rivals Open the Book of Life," *Daily Telegraph* (Feb. 12, 2001): 1.

267 "My Judeo-Christian heritage": Dugatkin 1999: 32–33, 173.

· *Bibliography* ·

Axelrod, R. 1984. *The Evolution of Cooperation.* New York: Basic Books.

Bateson, W. 1913. *Problems of Genetics.* New Haven: Yale University Press.

Boucher, D. H., ed. 1985. *The Biology of Mutualism.* New York: Oxford University Press.

Buchner, P. 1953. *Endosymbiosis of Animals with Plant Microorganisms.* English trans. 1965. New York: John Wylie & Sons.

Buss, D. M. 1994. *The Evolution of Desire: Strategies of Human Mating.* New York: Basic Books.

Carneiro, R. L., ed. 1967. *The Evolution of Society: Selections from Herbert Spencer's* Principles of Sociology.

Copeland, H. E. 1956. *The Classification of Lower Organisms.* Palo Alto: Pacific Books.

Crick, F. 1981. *Life Itself: Its Origin and Nature.* New York: Simon and Schuster.

Daintith, J., ed. 1994. *Bloomsbury Treasury of Quotations.* London: Bloomsbury.

Darwin, C. 1859. *The Origin of Species by Means of Natural Selection.* London: John Murray.

———. 1871. *The Descent of Man.* London: John Murray.

Davies, P. 1998. *The Fifth Miracle: The Search for the Origin and Meaning of Life.* London: Allen Lane: Penguin Press.

Dawkins, R. 1976. *The Selfish Gene.* New York: Oxford University Press.

———. 1986. *The Blind Watchmaker.* London: Penguin Books.

———. 1995. *River Out of Eden.* New York: Basic Books.

de Duve, C. 1996. *Vital Dust: Life as a Cosmic Imperative.* New York: Basic Books.

Desmond, A., and J. Moore. 1992. *Darwin.* London: Penguin Books.

Dickens, P. 2000. *Social Darwinism.* Philadelphia: Open University Press.

Douglas, A. E. 1994. *Symbiotic Interactions.* New York: Oxford University Press.

Dugatkin, L. 1999. *Cheating Monkeys and Citizen Bees.* Cambridge, Mass.: Harvard University Press.

Dyson, F. 1999. *Origins of Life.* Cambridge: Cambridge University Press.

Elliott, W. H., and D. C. Elliott. 2001. *Biochemistry and Molecular Biology.* New York: Oxford University Press.

Fortey, R. 1997. *Life: An Unauthorized Biography.* New York: HarperCollins.

Galton, F. 1883. *Inquiry into Human Faculty and Its Development.* New York: Macmillan.

Geddes, P., and J. A. Thomson. 1889. *The Evolution of Sex.* London: Walter Scott.

———. 1912. *Evolution.* London: Williams and Norgate.

Gilkey, L. 1985. *Creationism on Trial: Evolution and God at Little Rock.* Charlottesville: University Press of Virginia.

Gould, S. J. 1983. *Hen's Teeth and Horse's Toes.* New York: W. W. Norton.

———. 1989. *Wonderful Life.* New York: Penguin Books.

———. 1991. *Bully for Brontosaurus.* London: Hutchinson Radius.

Hawkins, M. 1997. *Social Darwinism in European and American Thought: Nature as Model and Nature as Threat.* Cambridge: Cambridge University Press.

Holland, H. D. 1984. *The Chemical Evolution of the Atmosphere and Oceans.* Princeton, N.J.: Princeton University Press.

Horrobin, D. 2001. *The Madness of Adam and Eve: How Schizophrenia Shaped Humanity.* London: Bantam Press.

Hoyle, F., and C. Wickramasinge. 1993. *Our Place in the Cosmos.* London: Orion Books.

Huxley, T. H. 1888. "The Struggle for Existence: A Programme." *Nineteenth Century* 132 (Feb.): 161–80.

———. 1897. *Evolution and Ethics.* New York: Appleton.

Jablonka, E., and M. J. Lamb. 1995. *Epigenetic Inheritance: The Lamarckian Dimension.* New York: Oxford University Press.

Jordanova, L. J. 1984. *Lamarck.* New York: Oxford University Press.

Kropotkin, P. 1902. *Mutual Aid: A Factor of Evolution.* Reprint, 1989, with an excellent introduction by George Woodcock. New York: Black Rose Books.

Kuhn, T. S. 1996. *The Structure of Scientific Revolutions,* 3rd ed. Chicago: University of Chicago Press.

Lovelock, J. 1979. *Gaia: A New Look at Life on Earth.* New York: Oxford University Press.

———. 1988. *The Ages of Gaia.* New York: Oxford University Press.

———. 2001. *Homage to Gaia: The Life of an Independent Scientist.* New York: Oxford University Press.

McMenamin, M. A. S. 1998. *The Garden of Ediacara.* New York: Columbia University Press.

McMenamin, M. A. S., and D. L. S. McMenamin. 1994. *Hypersea: Life on Land.* New York: Columbia University Press.

Margulis, L. 1970. *Origin of Eukaryotic Cells.* New Haven: Yale University Press.

———. 1981. *Symbiosis in Cell Evolution.* New York: W. H. Freeman.

———. 1997. *Microcosmos.* Berkeley: University of California Press.

———. 1998. *The Symbiotic Planet.* London: Weidenfield & Nicholson.

Margulis, L., and R. Fester, eds. 1991. *Symbiosis as a Source of Evolutionary Innovation.* Cambridge, Mass.: MIT Press.

Margulis, L., and K. V. Schwartz. 1998. *Five Kingdoms: An Illustrated Guide to the Phyla of Life on Earth,* 3rd ed. New York: W. H. Freeman.

Mayr, E. 1963. *Animal Species and Evolution.* Cambridge, Mass.: Harvard University Press.

———. 1991. *One Long Argument.* Cambridge, Mass.: Harvard University Press.

Mazumdar, P. M. H. 1992. *Eugenics, Human Genetics and Human Failings: The Eugenics Society, Its Sources and Its Critics in Britain.* New York: Routledge.

Monod, J. 1997. *Chance and Necessity.* London: Penguin Books.

Moore, J. R. 1979. *The Post-Darwinian Controversies.* Cambridge: Cambridge University Press.

Morse, S. S., ed. 1993. *Emerging Viruses.* New York: Oxford University Press.

Nutman, P. S., and B. Moss, eds. 1963. *Symbiotic Associations.* Thirteenth Symposium of the Society for General Microbiology, Royal Institution, London.

Passmore, R., and M. A. Eastwood. 1986. *Human Nutrition and Dietetics.* Edinburgh: Churchill Livingstone.

Pichot, A. 2000. *La Société Pure: De Darwin à Hitler.* Paris: Flammerion.

Portier, P. 1918. *Les Symbiotes.* Paris: Masson.

Ridley, M., ed. 1997. *Evolution.* New York: Oxford University Press.

Rose, M. R. 1998. *Darwin's Spectre.* Princeton, N.J.: Princeton University Press.

Ruse, M. 1979. *The Darwinian Revolution.* Chicago: University of Chicago Press. Reprint, 1999.

———. 1999. *Mystery of Mysteries: Is Evolution a Social Construction?* Cambridge, Mass.: Harvard University Press.

Ryan, F. 1993. *The Forgotten Plague.* New York: Little, Brown.

———. 1997. *Virus X.* New York: Little, Brown.

Ryan, F. X., ed. 2001. *Darwin's Impact: Social Evolution in America 1880–1920.* 3 vols. Bristol, Eng.: Thoemmes Press.

Sapp, J. 1994. *Evolution by Association.* New York: Oxford University Press.

Schneider, W. H. 1990. *Quality and Quantity: The Quest for Biological Regeneration in Twentieth-Century France.* Cambridge: Cambridge University Press.

Shermer, M. 2001. *The Borderlands of Science.* New York: Oxford University Press.

Singer, P. 1999. *A Darwinian Left: Politics, Evolution and Cooperation.* New Haven: Yale University Press.

Smith, J. M., and E. Szathmáry. 1995. *The Major Transitions in Evolution.* New York: Oxford University Press.

———. 1999. *The Origins of Life.* New York: Oxford University Press.

Sober, E., and D. S. Wilson. 1998. *Unto Others.* Cambridge, Mass.: Harvard University Press.

Stringer, S., and C. Gamble. 1994. *In Search of the Neanderthals.* London: Thames and Hudson.

Thomas, L. 1974. *The Lives of a Cell.* New York: Viking Press.

Vernadsky, V. 1997. *Biosphere.* Trans. David B. Langmuir, rev. and annotated by Mark A. S. McMenamin. New York: Copernicus Books.

Wallin, I. E. 1927. *Symbionticism and the Origin of Species.* London: Barriere, Tindall and Cox.

Weinberg, S. 1999. *A Fish Caught in Time.* London: Fourth Estate.

Weindling, P. 1989. *Health, Race and German Politics Between National Unification and Nazism, 1870–1945.* Cambridge: Cambridge University Press.

Wilson, E. O. 1975. *Sociobiology,* abridged. Cambridge, Mass.: Belknap Press of Harvard University Press.

———. 1978. *On Human Nature.* Cambridge, Mass.: Harvard University Press.

———. 1992. *The Diversity of Life.* London: Penguin Books.

Wynne-Edwards, V. C. 1962. *Animal Dispersion in Relation to Social Behaviour.* Edinburgh: Oliver & Boyd.

· *Glossary* ·

Adaptation: a feature of an organism that is useful for survival.

Allele: a gene variation that occupies a given position on a chromosome and usually codes for a variation of a particular protein.

Allopatric speciation: the origin of a new species from an existing one through geographic isolation.

Amino acid: the smallest chemical building block of a protein. There are twenty in all, each a compound of carbon, hydrogen, oxygen, and nitrogen.

Archaea: the earliest forms of cellular life, dating from before the oxygenation of the atmosphere. According to Carl Woese, they have distinctive ribosome RNA structures and other chemical features. These life forms include methane-producing bacteria, halophiles (bacteria that require salt water), acid-lovers, and bacteria that metabolize sulfur in hot-spring habitats. Also called archaebacteria.

Archaebacteria: see **Archaea.**

Australopithecus: a genus of hominids that inhabited Africa as early as 4.4 million years ago. Their brains were midway in size between those of humans and apes. Most australopithecines lacked tools but could walk upright.

Benthos: the collection of marine life that inhabits the bottom of an ocean or lake. The adjective form is benthic.

Binary fission: the process of cell division by which bacteria and some protists reproduce.

Black smoker: a thermal vent found in the deep oceans along fault lines. Smokers exude sulfurous compounds that provide energy for symbiotic microbes.

Blue-green algae: a term formerly used to describe the photosynthetic life forms that are now classed as cyanobacteria.

Chloroplast: the organelle within the cytoplasm of plant cells that enables photosynthesis to take place. See also **Plastid.**

Cyanobacteria: see **Blue-green algae.**

Diffuse symbiosis: a type of exosymbiosis in which the interacting partners do not physically meet but share metabolic products through air, water, or another such medium.

Diploid: containing two copies of each chromosome in the nucleus.

DNA: deoxyribonucleic acid, the molecular basis of genes and chromosomes in all life forms except some RNA-based viruses.

Ecology: the study of the distribution and abundance of organisms in their natural environment.

Ecosystem: all of the organisms that live in a particular ecological niche, together with the physical environment.

Endosymbiosis: a symbiosis involving the union of different genomes, at either the nuclear or the cytoplasmic level.

Eon: the largest formal unit of geological time; for example, the Archaean.

Era: a division of geologic time shorter than an eon but longer than a period; for example, the Paleozoic.

Eubacteria: all bacteria other than the Archaea.

Eukaryote: a life form consisting of a cell or cells in which the cytoplasm is separated from a distinct, membrane-bound nucleus. All organisms on Earth other than bacteria are eukaryotes.

Exosymbiosis: all forms of symbiosis other than endosymbiosis.

Fitness: the measure of an individual life form's probability of survival and, thus, of reproductive success.

Gamete: the male or female sex cell in a plant or animal. In humans the gametes are the sperm and egg, each containing twenty-three chromosomes.

Gene: the DNA (or, in some viruses, RNA) coding for a single hereditary unit, usually a protein.

Genome: the entire genetic makeup of an organism or species. The human genome comprises our forty-six nuclear chromosomes, together with the mitochondria and other structures in the cytoplasm.

Genotype: the genome of an individual life form.

Haploid: having a single copy of each chromosome in the nucleus.

Herbivore: an animal that feeds on plants.

Holobiont: the entire partnership in a symbiosis; see **Mycorrhiza,** for example.

Host: the larger partner in the commonest form of symbiosis. See also **Symbiont.**

Macrosymbiosis: a diffuse symbiosis on a global scale; for example, in the oxygen and carbon dioxide cycles, plants produce oxygen, which is taken up for respiration by animals, which expire carbon dioxide, which in turn is taken up by plants.

Meiosis: the reduction division of sex cells that results in the formation of germ cells, or gametes, which are haploid (contain only a single copy of each chromosome). As part of fertilization, the two gametes combine to form the fully diploid zygote, which now contains matching sets of chromosomes (in humans, forty-six) from each parent.

Mitochondria: Organelles, originally derived from symbiotic bacteria, found in the cytoplasm of oxygen-breathing cells.

Mitosis: asexual reproduction (cell division) of eukaryotic cells, including protoctists.

Multicellular: having many cells, as applied to the organization of a fungus, animal, or plant.

Mutation: a random change in the nucleotide sequence of a gene.

Mycorrhiza: a symbiotic association between plant roots and fungi, whereby the fungi greatly increase the roots' absorptive capacity within the soil and the plant supplies the fungus with nutrients.

Natural selection: the Darwinian concept that nature selects, among the members of a species, those individuals best adapted for survival.

Ontogeny: the sequence of development of an individual organism from conception, through embryological development and subsequent life, to death.

Organism: life form.

Paleontology: the study of ancient life, based largely on fossils.

Parasite: an organism that lives in or on another life form and draws its nutrition directly from it.

Phenotype: the observable physical characteristics of an individual in terms of body, tissues, biochemistry, and physiology. Compare **genotype.**

Phylogeny: the evolution of a genetically related group of organisms, often expressed as a schematic drawing of that branch of the evolutionary tree.

Plasmid: a discrete genetic element that can move between genomes.

Plastids: photosynthetic organelles in the cells of algae, plants, and animals that were originally derived from a variety of life forms. They include chloroplasts.

Polyploid: having more than two copies of the chromosomes in the nucleus.

Prokaryote: a life form, usually a bacterium, with a simpler structure than a eukaryote.

Protein: a molecule made up of many amino acids. It is coded within the genome by one or more genes and plays an important part in the structure and function of biological bodies.

Protists: unicellular eukaryotes. In Margulis's classification, protists are the smallest of the protoctists. See **Protoctista.**

Protoctista: a kingdom of diverse life forms that includes what were previously called the Protozoa. Though neither plant nor animal, some protoctists gave rise to all plants, animals, and fungi. They are eukaryotic (nucleated), in contrast to bacteria, and may consist of a single cell (for example, the amoeba) or multicellular forms, such as seaweeds, which are algae. See **Protists.**

Protozoan: a term formerly used to describe some protists.

Recombination: the swapping of matching segments of chromosomes during the first stage of meiotic reproduction.

Rhizobia: nitrogen-fixing bacteria that live in the root nodules of legumes.

RNA: ribonucleic acid, the molecular basis of genes in RNA-based viruses and the basis of a number of different genetic messenger and production molecules that relay the coded information of nuclear DNA to the protein factories, called ribosomes, in the cytoplasm of the eukaryotic cell.

Sedimentary rock: rock consolidated by the accumulation of loose sand or sediment, or by precipitation from a watery solution.

Seed: the reproductive structure of plants, resulting from the union of male and female gametes.

Speciation: the origin of a new species from one or more that already exist.

Species: a group of individuals that can interbreed and produce potentially fertile offspring.

Spore: a reproductive propagule of bacteria, fungi, and some plants that is distinct from a zygote or seed. It is often resistant to heat and desiccation.

Stromatolite: a layered structure of carbonate or silicate rock, produced by the chemical and metabolic actions of communities of microorganisms, including cyanobacteria.

Symbiont: one of the partners in a symbiotic relationship.

Symbiosis: As originally defined by de Bary, an association between differently named species that persists for a long period.

Symbiogenesis: evolutionary change as a result of symbiosis.

Zygote: the fertilized cell that results from the union of the male and female gametes during reproduction.

· *Index* ·

7 6/07

Consider this my little revenge on the Minnesota Aetheist Assosiation Meeting I attended recently. ~~in which a proff~~

Evolution was the subject and ever the boast. They claimed to take delight in ridiculing creationists.

Evolution will not explain away God, so people can go on living in their sins and not be accountable to moral decency.